PLASTICS: JUST A LOAD OF RUBBISH?

PLASTICS

JUST A LOAD OF RUBBISH?

Re-evaluating Plastic and Its Role
in Saving the Environment

Alicia Chrysostomou

HERO, AN IMPRINT OF LEGEND TIMES GROUP LTD
51 Gower Street
London WC1E 6HJ
United Kingdom
www.hero-press.com

First published in English by Hero in 2023

The right of Alicia Chrysostomou to be identified as the author of this work has been asserted in accordance with the Copyright, Designs and Patents Act 1988. British Library Cataloguing in Publication Data available.

Cover design by Ditte Løkkegaard

Printed and bound by in the UK by Severn Print

ISBN: 978-1-91564-379-7

To Sebastian,
for whose generation we need
to get things right

Contents

Preface

Plastic fantastic or just a load of rubbish? Plastic demands our attention in a way few other materials can manage. It has risen and fallen in our estimation to a spectacular extent in the last few decades. Our sentiments towards plastic now have probably hit rock bottom. But why?

Just before the pandemic encircled the globe, I found myself in a lecture theatre surrounded by a hundred or so very excited nine-year-olds as they watched a chemistry professor demonstrate the properties of gases. There were exploding balloons, thick clouds of dry ice and gushing foams. The children were enthralled. The professor explained that helium was now a valuable gas and in consequence some retail outlets offered children promotional balloons on the end of sticks rather than the traditional floating gas-filled alternatives. He leaned forward and urged the children not to take these bobbing gifts. The sticks are made of plastic, he told them. Then added, 'And what happens to the plastic?' The children chanted back as one, as if reciting a well-rehearsed mantra, 'It ends up in the sea.'

I question the wisdom of telling children that plastic, as a matter of course, ends up in the sea. Plastic only ends up in the sea if it's put there. It doesn't sprout legs and head straight to the ocean the moment its usefulness expires. Or are there lines of children standing on clifftops flinging their unwanted plastics into the sea? Of course not.

Plastics waste is a major cause of concern, yet for the most part this waste is correctly disposed of and much is recovered

I

and recycled. There is no denying that this is not always the case, particularly if we look globally, and this needs addressing. But even where the best intentions are meant, confusion still arises.

Another misleading credential associated with plastics, and now seen with increasing abundance, relates to bioplastic. If you automatically heave a sigh of relief when you see this stamped on a product and think, 'Well that's all right then,' all may not be as it seems. The 'bio' here most often does not mean biodegradable: it just refers to the biological origins of the plastic. In so many cases when a plant is converted into a plastic, it is just that: a plastic just like its oil-derived cousin with all the inherent pros and cons of that material. This is a whole other area of confusion that only serves to alienate the public yet further from the attributes of plastic and will be looked at in this book.

The key underlying question to ask is whether this material is as bad as it has been painted. Based on my professional experience as a polymer scientist and engineer, I would argue it is not.

So for those of you who automatically consider plastics as being the scourge of modern society, I'm afraid this might not be the book you thought it was. I'm giving fair warning, this is not a plastics-bashing book. I know I'm going against the norm but here's the challenge: can you give the other side a fair hearing? The result might be surprising.

I'm certainly not intending to utterly exonerate all uses of this valuable material. Serious problems undoubtedly exist, but I would like to redress the balance and ask if it's right to heap the ills of the planet on this one class of material. Worse, by fixating on plastics are we actually taking our eye off the ball and not seeing other areas worthy of equal or greater concern? Might we even consider the possibility that we may be able to rectify at least some of our problems with the assistance of plastic? The Covid-19 pandemic is a good example, with vast quantities of equipment used against this virus – from PPE to test-kit components, from breathing tubes to ventilator parts – all being made from plastic, moreover that bane of society, single use plastic.

Many column inches have gone to declaim the depletion of non-renewable fossil fuels –especially oil – in the manufacture of plastics. Hang on a minute though; just how much oil is actually being depleted in the name of plastics? The answer may surprise. Globally, just 4% to 6% of oil produced goes into the manufacture of plastics. So what happens to the rest? Largely it's incinerated. The vast majority of extracted oil is burnt for fuel; there is no going back on this finite resource. By converting it into plastic it becomes something tangible, useful and recyclable.

It is interesting to note that concerns over plastics use are not by any means new or unique to the current age. Previous generations saw that limits were needed on environmental grounds. A textbook[1] written by the highly respected academic and plastics expert Dr John Bryson throws an interesting light on early concerns. The following excerpt, taken from a passage in its introduction, comes from a revised 1977 edition:

'Another problem confronting the plastics industry, and in fact civilisation as a whole, which came to the foreground in the 1970s was the concern for the environment. There has been increasing awareness of the need for conservation of resources, and of the evils of pollution and of the fact that it is the quality of life rather than material possessions which is the criterion by which civilisation should be judged.'

It is interesting to note that these concerns have been around for nearly fifty years and continue to be expressed. Might it be that our desire for material possessions hinders our actions when it comes rectifying harm done to the environment? We do however need to acknowledge the benefits of plastics to our quality of life. Somewhere a balance needs to be made.

Plastics have been instrumental in giving us healthy, fuller and more socially inclusive lives. We can do things we never would have imagined just a couple of generations ago. Our lives have improved in so many ways; we have choices in how we live, what we wear, what we eat and even what leisure activities are open to us. Now to ask the important question, how much of this do we want to give up, how much *should* we give up and what would be the consequences to the planet if *all* plastics were banished? And come to that, why do we have such strong opinions about plastics in the first place? I will try to address these issues and others in the pages that follow.

Let's start with the question, why have we formed such strong opinions about plastics?

1

WHY ARE WE SO SET AGAINST PLASTIC? THE INFLUENCE OF THE MEDIA

There have been some fantastic programmes on the television these last few years. We have been entertained and enlightened on a terrific range of subjects. Some of them have become seminal and have impacted society in ways not imagined. One such was the excellent BBC offering; *The Blue Planet*. It became a global sensation, and the episode in its second season focusing on plastics spawned international debate, kickstarting a multitude of campaign groups, calls for action and even influenced governmental policy. It has helped shape the thinking of a whole generation.

The phenomenal success of this programme led to a plethora of others, all determined to pick up the environmental mantle and in doing so, increasingly demonise plastics. After all, this is what the public wanted to hear. Plastics are the root of all evil. They are clogging the sewers and abound in the oceans, congregating in sprawling islands of waste. The media, in its bid for ever more sensationalist headlines, ramp up their efforts to further highlight the scourge that is plastics.

Before having a look at specific myths about plastic, we need to figure out why plastic took such a foothold in the first place. Why is it so omnipresent? Do we need to be so totally dependent on this material and is it being used to its best effect? And probably the biggest question, what on earth, and I use that word advisedly, is such vast quantities of the stuff doing in the seas and oceans? Once we figure that out we can consider the need (or otherwise) to contain its use and whether we really do need an outright ban on all things plastic.

It will help if we go back to the beginning, look at the origins of this material and discover how it gained such dominance in just a few decades. This potted history might show how essential these materials were, even right at the start, in conserving the natural environment. It will also show how much came about accidentally with many breakthroughs resulting from spills, mishaps and even, on one occasion, a clumsy cat prowling a lab.

2

A SHORT HISTORY –
WHERE DID IT ALL START?

Plastics really came into their own in the twentieth century with developments in polymer[2] chemistry and an understanding of material synthesis, although they really began their ascent a few decades earlier in the nineteenth century. These early plastics were bioplastics, although this wasn't how they were defined back then. Then they were just plastics in the same way that our telephones were telephones and not land lines, and mail came through the door with no snail involvement, figuratively or otherwise.

The earlier years of plastic, or polymer, development focused on materials derived from natural sources, while the latter part of the twentieth century saw great advances in the development of highly sophisticated synthetic plastics with a broad range of attributes. The first fifty years or so saw the introduction of barely a dozen plastics, but the1950s really was the turning point of plastic's widespread adaptation with both an intense flurry of discoveries and usage rocketing from that point.

Rubber bounces into our life

The modern story of polymers probably starts when a useful form of rubber was developed in the 1840s. This coincided with the discovery of vulcanisation. Rubber of course existed and was used for centuries in Europe and millennia in its native South

America. The breakthrough discovery came when it was found that the addition of sulphur could act as a vulcanising, or hardening, agent and this really expanded the usefulness of the material.

It took another leap forward during the Second World War when major logistical problems in supply meant synthetic alternatives to natural rubber had to be found. Rubber was an integral part of military vehicle tyres so an alternative had to be found if natural sources were unreliable.

Once development of a synthetic rubber for tyre manufacture was established, other synthetic rubbers followed. We now have quite a variety of rubbers suitable for all sorts of uses, from wetsuits to seals, window beading to skyscraper earthquake dampers. Not only are there a number of different types of rubber now in existence, they can also be modified and adapted for an even broader range of uses. For instance, a wetsuit would be manufactured from a particular rubber called neoprene (more properly known as polychloroprene), but adding certain fillers makes it a suitable material for making airport baggage-handling conveyer belts. Pretty much all the varieties of rubber now available can be adapted for specific applications using an often complex selection of fillers and additives.

One defining and fundamental feature of rubber, certainly in its early days, centred on its structure. Rubber tended to be lightly cross-linked – in other words there was light bonding between the internal molecular chains forming its structure. This cross-linking helped give rubber its characteristic stretchiness, but did mean that it couldn't be conventionally recycled. In other words, it couldn't be melted down and formed into something else. Now however, some rubbers can do just that. They have been developed to allow recycling and reforming post-use. They are called thermoplastic rubbers, or thermoplastic elastomers, and straddle the camp between the previously distinct groups that could be classified as rubber and plastic. This is why it is probably better convention to just call all these materials polymers – a more catch-all term that takes account of the overlapping that exists.

Gutta-percha

Before moving on to plastics it is well worth mentioning another rubber-like material, Gutta-percha. It is closely related to the more familiar natural rubber, although not as elastic in nature, and came to prominence at around the same time. Gutta-percha comes from the dried sap of the 'percha' tree indigenous to South East Asia. It was 'discovered' by an English explorer in the 1650s, but was of course known of and used by native tribes for hundreds of years previously. It was then considered somewhat of a novelty material, but it really came into its own on an industrial scale in the 1860s, when another Englishman realised its potential, initially as a medical material. It made an appearance in the Great Exhibition of 1851, where its wide-ranging capabilities were showcased.

Indeed, gutta-percha is a remarkable material and can be said to have kickstarted the global telecommunications age. In the 1860s a scheme to lay a transatlantic cable enabling communication across the Atlantic was formulated. It was soon discovered that gutta-percha made an excellent electrical insulator that didn't disintegrate in seawater. It was soon sheathing the many thousands of miles of cables criss-crossing seabeds around the world. Before that time messages took prodigious lengths of time to arrive at their destinations. If those messages were sent from India or Australia, the wait was immense. Laying the cables allowed coded messages to click around the globe practically instantaneously. It made global communications possible in a way they had never been previously. Now messages could be received and responded to immediately.

Gutta-percha was found to have yet more uses aided by its happy ability to mould easily. It just needed a little heat (around 70°C) to become pliable, after which time it could be moulded and allowed to set hard by cooling to room temperature. It soon found uses in the medical sector, dentistry (where it was welcomed as a form of tooth filling), garments, musical instruments, jewellery and for making golf balls – a major use in the late nineteenth century.

Early extraction methods meant felling the trees to recover the latex. With high demand from the West, this meant many hundreds of thousands were lost, leading to the near extinction of this species of tree. Extraction techniques have fortunately changed, and now the latex is extracted by incising the bark (as with natural rubber) or crushing the leaves, and so gutta-percha remains with us today, albeit used on a much smaller scale than in its nineteenth-century heyday. Now its application[3] lies primarily in dentistry, particularly endodontics, although it can still be found in golf ball casings.

Plastic firsts

Plastics in their truly definitive sense have been around a long time. A plastic material refers to any material that can be softened by heat and then shaped. At its most basic it is a material that is malleable. Of course the first of these plastic materials had completely natural origins and needed very little other than the application of heat to form them into a useful product.

A material that fulfils these properties and that could be defined as a plastic is amber. It was well-known in antiquity and has been used for millennia. It is a naturally polymerised fossilised resin that has an ability to be shaped and formed once it reaches a certain temperature. Its uses tend to be predominantly in the field of jewellery, although it has been used in other applications, particularly for ornaments and smoking and writing paraphernalia.

Tortoiseshell has also been used since antiquity for all manner of applications, from furniture inlays to musical instruments.

It continued to be used over the centuries and was particularly popular with the Victorians, when it was commonly found in more affluent households gracing the dressing tables of ladies of taste and distinction in the form of combs, brush backings and hair adornments. It remained in use well into the twentieth century, in applications such as spectacle frames and guitar picks, which are now, thankfully, illegal.

Shellac, again a natural material known for millennia, has also endured into modern usage. Totally natural in origin, it actually comes from an excretion made by lac beetles commonly found in the Indian subcontinent. Originally it was used as a dyestuff and lacquer before becoming the material of choice for the manufacture of gramophone records. Mixing it with wood flour meant it could be moulded into all sorts of decorative items, so extending its usefulness.

It is still very much in use today, with its virtues seen emblazoned on window posters of nail bars across the country, where it is used as a hard lacquer coating for nails. But it has far more uses than that, uses that are not immediately apparent to the consumer. It also turns up on food ingredient lists as E904, used as a fruit and sweet glaze. So if you wonder how some sweets and chocolates get their very shiny appearance and don't readily melt in your hand, look at the labelling. If E904 is mentioned then you know your sweets are shining with the gleam of shellac armour.

Not labelled on the product but just as shiny is the fruit on display in many supermarkets. Fruit is washed before appearing on our shelves, which removes its natural shine, so a replacement 'wax' is used. A number of materials can be used for this function, but shellac is still often to be found shining an apple or welling the dimples of an orange.

Shellac is very versatile. It can just as easily be found as a hardener for the points of ballet shoes or as a waterproofing lacquer sold in DIY stores. Mention will be made again of this interesting material when we consider the ethical uses of some natural materials.

Before that, another natural plastic material worth a mention is horn. Animal horn was scraped and sliced into thin sheets then heated to flatten. Once flat it was placed into a square frame and lit from within, making a lanthorn or what we now more commonly call a lantern. This material was also used for centuries for a variety of applications including buckles, beakers and buttons as well of course as shoe horns. Taking the lantern example, the horn, like the shellac and tortoiseshell, was used to create a new product. However, materials now began to be developed that could replace other more costly or rarer originals.

Accidental discoveries happen with surprising regularity. So no surprise that the next milestone in plastic material development began as an accident. We have now come to cellulose, the precursor to celluloid. Cellulose can be derived from wood or cotton, and when it is mixed with nitric acid, it can produce some surprising reactions. A German-Swiss chemist by the name of Christian Friedrich Schönbein experimented with these materials in the mid-nineteenth century. He managed to spill some on his wife's apron which he happened to be wearing at the time. After hanging it out to dry he found that the apron had gained explosive qualities when touched. He had accidently made cellulose nitrate, known at the time as gun cotton. This went on to become cordite in the UK.

This material came to the attention of Alexander Parkes, who altered the ratios of cellulose to acid and found he could make something really useful. This new material became a true plastic. Parkes then experimented by adding combinations of other chemical ingredients such as camphor, a move which proved fortuitous in extending the peripheries of material science. Parkes managed to produce a material that was tough and elastic and, importantly, could be moulded and shaped. This material became known as Parkesine and is widely accepted as the world's first man-made plastic.

It was showcased at the 1862 International Exhibition, where it was welcomed as a material that could be substituted for the more expensive and increasingly rarer ivory and tortoiseshell. This substitute material facilitated the development of consumerism, offering otherwise exclusive products to a mass market.

Parkes was the first to exploit this new plastic material commercially, but he was no businessman, and eventually his Parkesine Company folded. A collaborator, Daniel Spill, went on to set up a new company called the British Xylonite Company making Xylonite and Ivorite. This company managed to tap into the new market for detachable washable collars and cuffs, among other products, meaning these frequently grubby items could be given a quick wipe clean without having to launder the whole shirt. The company was bought and sold many times over the years, becoming an ever smaller part of larger companies, but it does survive today.

Crossing to America, Parkesine was taken up and further adapted by the Hyatt brothers, who called their material Celluloid. They wanted just such a material to enter a competition that sought a synthetic alternative to ivory for billiard balls. They originally devised a method of making these from cloth, ivory dust and shellac, coated with celluloid. This worked well, although an interesting side effect emerged. Celluloid is explosive and inflammable when impacted with force (remember its gun-cotton origins). A letter sent by a saloon owner in 1869 tells of an unusual outcome of this unfortunate reaction. He noted that when his costumers knocked billiard balls with force they would produce a mild explosion, at which point all the men in the room would pull their guns!

The Hyatt brothers continued their development of Celluloid. They saw how important camphor was to the process and this was the breakthrough needed to make it a commercial possibility, something that eluded Parkes. They also sorted out the volatility of the material and, importantly, developed an injection moulding machine that could shape the material. Now it really took off making items as diverse as dentures and dolls, fountain pens and spectacle frames. It could even be used as a replacement for baleen, a whale body part used in the manufacture of stays for corsets.

Cellulose nitrate is still used today, notably to make ping-pong balls. Although less explosive than they once were, they will still ignite quickly and burn vigorously.

Next on our list is casein, another example of a material that was developed as a cheap alternative to costlier traditional materials. It too is a semi-synthetic material, i.e. it has a natural source but also the handy ability to be processed into a longer-lasting malleable material.

Casein is derived from milk protein. In fact, its journey in conversion to a plastic material starts not unlike the process used to make cheese (the name is derived from the Latin word for cheese). The milk protein needs to be extracted and this is achieved by warming milk. Then an ingredient is added which ensures that the milk splits into curds and whey (vinegar easily does the trick). The whey is flushed out and the remaining curd is the casein protein.

Casein was known for a great many years in its unhardened form, which is relatively weak, with the Ancient Egyptians using it as a pigment fixative in their wall paintings. It continued being used as a glue for millennia until it came to the attention of a nineteenth-century German scientist, Adolf Spitteler, who developed the hardened form, enabling more widespread use, e.g. in the manufacture of products such as knitting needles, buttons, buckles and umbrella handles. Strictly speaking it was

Spitteler's cat who accidently hit on the solution, knocking over a bottle of formaldehyde in the lab. The formaldehyde spent the night dripping into the cat's bowl of milk, which by morning had curdled and set firm. Spitteler found the horn-like material that had coagulated and it set him thinking. By 1899 Spitteler, collaborating with a printer by the name of Wilhelm Krische, had set up industrial-scale production and had taken out a patent for a washable white casein 'slate' for children to practise their writing (an early version of a whiteboard). The material continued to be adapted and did very well as a substitute material for bone or horn used at the time to make buttons. Casein buttons were found to be able withstand the touch of a hot iron, and better still could be washed and even dry-cleaned with impunity. Great news for the laundress who would otherwise have to remove and resew buttons each wash-day. Applications for casein continued for decades, although now it has fallen mostly into disuse.

Next to enter the arena is the world's first fully synthetic material (those coming earlier used natural materials as a starting point). A Belgian chemist called Leo Baekeland began working on phenol and formaldehyde, two chemicals known to have interesting effects under certain conditions. Baekeland combined these chemicals and then, experimenting by adjusting temperatures and pressures found, he had made a curious mouldable resin-like material. He named this Bakelite.

By 1910 Bakelite really took off. It had a tendency to be brittle so needed to be filled. Baekeland found that mixing his plastic with fibres such as those of wood, asbestos or cotton gave the desired results and solved his brittleness problems. The material could now be moulded into a number of products. Initially it was used primarily in electrical insulation, then also an emerging industry. It grew hugely in popularity and went on to be used in jewellery, combs, radio and telephone cases, and a vast array of other products – just as long as they were brown!

The phenol (a.k.a. carbolic acid) Baekeland used was derived from coal tar. This was an interesting by-product of coal gas, which itself had been developed over the previous century or so for providing light and fuel. Coal tar was incidental to the production of the gas but found to be very useful in its own right. This was demonstrated in another fortunate accident when a barrel of coal tar spilt onto a macadamed road (named after John McAdam, who devised a method of surfacing roads with aggregate so making carriage travel easier). The potential for sealed roadways was spotted immediately and tarmacadam, or tarmac, was born. Charles Macintosh also foresaw the potential of this tar and produced another by-product called naphtha, which he used in conjunction with rubber to rubberise fabric – and now the waterproof 'mac' was born.

Coal tar was actually found to be a valuable material and found its way into a vast array of products. From this otherwise waste material, a number of unexpected applications were found including soap (think carbolic), creosote, shampoo, antiseptic, saccharine and paracetamol, to name just a few. Many are now derived from synthetic sources, but paracetamol can still be made from coal tar. Creosote has been banned since 2003 because of its carcinogenic properties, as has carbolic soap, although for many the pungent scent of each is easily evoked.

Coal tar as a by-product of gas was soon on its way out, as eventually coal gas gave way to town gas (primarily produced from coal but could be produced from other sources such as oil derived naphtha), and this yielded in turn to North Sea gas. Less reliance on coal gas led to a shift from coal tar being the principal source for naphtha. Luckily this came at a fortunate point in history. It coincided with the ascendance of the motor car and now that other major fossil fuel was needed. That, of course, was oil.

CHAPTER 2

Development of synthetic plastics

Oil was required in ever-increasing quantities to power the modern world. However, to be made useful it needed refining. This is done through a series of fractions and distillations. Fractioning the oil enables the development of a number of products, first the heavier bitumen and tars through various oils, used for heating, marine transport and so on. Next come the diesels, and then naphtha. Naphtha is 'cracked' to give petrol and the by-product ethylene. In effect, this is the dregs of the oil refining stages, but it didn't take long for the enquiring minds of the day to find a use for this by-product, which speaks volumes of the no-waste approach then prevalent.

It was soon found that mixing this ethylene – which is basically just bundles of carbon and hydrogen – with chlorine gave some very interesting results, and in 1912 the beginnings of PVC, or polyvinyl chloride, took root. Actually that was the third 'beginning' of PVC, it was initially catalogued in 1872, having been first observed in 1838. But no one really noticed the potential of (or even initially patented) this nascent PVC until it was resurrected yet again, and from 1933 it really took off. It's now one of the most commonly used plastics on the market.

PVC was the first plastic derived from the ethylene extracted from oil, but what about the chlorine bit? That actually comes from salt. (The vinyl part of the name just means that the ethylene bundle is missing one hydrogen atom, and that is where the chlorine slips in.) Approximately 57% of PVC has its origins in salt so in reality uses comparatively little hydrocarbon, certainly compared to other plastics which came hot on its tail.

Experimentation continued and reaped rewards. Soon PMMA, a.k.a. acrylic, came into being, then the workable form of PVC mentioned, then nylon, polyethylene and, on the cusp of the

Second World War, that packaging favourite, PET. These materials really showed their capabilities during the war, particularly ones like acrylic that could be used instead of glass for aircraft canopies, nylon for parachutes instead of silk, and fibreglass encased within epoxy for aircraft radomes.

It may be interesting to dig a little deeper and uncover just how these materials were discovered. Needless to say happy accidents pepper their stories. These plastics will crop up many times in the following pages, sometimes adding their scientific names to common parlance (polythene, which is a contraction of polyethylene), some better known by tradenames rather like the Hoover of the plastics world (Teflon), and others whose names are such a mouthful that it is best to leave them in abbreviated form (Teflon is a tradename for PTFE, a.k.a. Polytetrafluoroethylene).

More discoveries and more accidents

As mentioned above, PVC passed in and out of history twice before its true potential was properly realised, starting back in 1838 with a French scientist by the name of Henri Victor Regnault, Regnault was a terrific experimentalist and worked prodigiously in developing vinyl chloride, a compound that at the time had important uses industrially. While working on the gaseous compound, Regnault stoppered some into a glass tube and left it on a shelf. Some days later he happened to revisit the shelf and found the vinyl chloride had polymerised thanks to its exposure to sunlight, making polyvinylchloride, a.k.a. PVC. In polymerised form this PVC just looked like a white powdery substance; Regnault had no idea what this was or how it could be used, so he wrote it up and left it shelved. He didn't even bother to publish a paper introducing this new material until in 1872 another scientist also happened to stumble across it. This time it fell to German chemist Eugen Baumann to rediscover PVC. Baumann started his working life as an apothecary apprenticed to his father. He passed his pharmacist's exams and took on a doctorate, writing a dissertation on vinyl compounds. He continued experimenting on these compounds when in 1872 he again stumbled on a powder that was PVC after leaving a flask of the base material in the sun. It was again shelved as an odd material, difficult to work with and with no clear applications. Fast forward to 1913 when a German inventor patented PVC polymerised naturally with sunlight. Again there was no real use attached to the novel material, just an acknowledgement of its existence.

The real breakthrough came about in the 1920s, when an American chemist called Waldo Semon again came across the material, but this time realised that it could effectively be used as a rubber replacement for waterproofing fabrics. Semon, true to form, made his discovery of PVC by accident. He had been tasked with finding a method for bonding rubber to metal and, having run out of rubber to experiment with, tried producing it synthetically. Remember, PVC as a substance was known; its value or use was not. Semon experimented by heating PVC powder in solvents, thus providing the breakthrough needed to unlock its potential. Its early applications lay particularly in the area of waterproof coatings, but it also found use as shoe heels, shower curtains, umbrella fabrics, records and even garden hoses. By the Second World War its potential was realised again as wire insulation on military ships. From there its usefulness accelerated and can now be found in many facets of life from flooring to pipes, electrical insulation to fabric coatings in addition to so many of its original uses. Crafters too have discovered its benefits in the easily mouldable polymer clay which is also PVC based. It is now one of the most widely used plastics on the market and has come a long way from sitting on a shelf waiting for a spot of sunshine.

Our next material is nylon. One immediately apparent aspect of this material that sets it aside from many other plastics is its name. So far the prefix 'poly' has been used for each of the synthetic materials considered here, so it may be no surprise that nylon too has a name that follows this format: scientifically nylon is a polyamide (PA). Nylon is lauded as the material that marked the turning point in the syntheses of plastics. It was discovered in America by a troubled soul called Wallace Carothers. Carothers was a chemist working for the leviathan that was DuPont in the 1920s (although he began his studies following his father's footsteps in the field of accountancy before becoming a lecturer in chemistry at Harvard). He began his research while at university, becoming interested in the field of polymers, then very much in its infancy.

Although most famous for his work on nylon, Carothers was also associated with the discovery of another well-known material, and that is neoprene. This was the first breakthrough material involving Carothers, whose boss was then on the hunt for a synthetic rubber. It wasn't long before his colleague, Arthur Collins, found the solution needed: he dabbled at distilling some experimental reactions and ended up with a liquid of unknown composition. Not sure what to do with it, he put the liquid aside in a stoppered tube and, revisiting it some days later, found the now hardened contents bounced. So in 1930 neoprene was synthesised, becoming the first commercially successful speciality rubber. Neoprene was and still remains a terrific alternative to natural rubber, although their properties do differ, and is used everywhere from wetsuits to seals, tubes and even orthopaedic supports. Neoprene is a tradename (belonging to DuPont) and in actuality is a polychloroprene. On the back of this discovery, Carothers continued his research into elastic fibres with mixed success. The stumbling block was finding a fibre capable of withstanding chemical attack during the dry-cleaning process.

Carothers was disheartened at his lack of progress, but his grasp of the new discipline of polymer science stood him in good stead. He was one of the first to understand the role molecules played in building up these new synthetic (and indeed natural) polymers. He found a way to build up molecules through a series of reactions and discovered these reactions could produce novel materials. He also found he could string together long molecules rather like linking together a chain of paper clips. The point was that they could hook on to each other and this formed an important stepping stone in the understanding of how polymers could be constructed. It was a major breakthrough in understanding how these materials worked and how new ones could be synthesised from disparate compounds.

His goal though was to find a synthetic material that could be used as a textile fibre, and this was to prove elusive for several years. Another colleague by the name of Julian Hill then discovered polyester fibres, but at this time they weren't of much use commercially as they broke up in boiling water. After a great deal of hesitancy with work stopping and starting, Carothers, in association with Hill, turned to a group of chemical compounds called diamines and ploughed on. Finally in 1935 a workable polyamide was produced; moreover it could form a fibre, had a degree of elasticity and could be commercially processed. The research into this elusive material had cost DuPont tens of millions of dollars to develop, but it took a worse toll on Wallace Carothers. He had suffered from severe bouts of depression throughout his life and in 1937 sadly committed suicide. After overcoming each taxing problem and hurdle throughout the development of nylon, he died just when the material exploded with great success onto the waiting world. In 1938 DuPont built a factory to manufacture nylon fibre and in 1939 industrial production started. It was so revolutionary in the field of women's hosiery that the word 'nylons' is still synonymous for stockings. During the war years (in as far as the USA was involved), all operation was diverted to military use, and so nylon parachutes evolved. Post-war the use of nylon massively expanded from a multitude of textile uses, including rope-making, to bristles for toothbrushes and all those other moulded internal gadgets, gizmos and cogs needed in everyday life. Nylon is still a vital part of our world today.

But where did the word nylon come from? There are many common myths, and even DuPont themselves appear to have released conflicting accounts of its origin, but the most likely source appears to have been Ernest Knight Gladding, who headed a department within the company. Several names had been toyed with including 'Wacara', a contraction of Wallace Carothers, and 'Delawear', which was a play on the fact the new nylon factory was sited in Delaware. Gladding then suggested 'Norun' to emphasis the aspiring qualities of the textile when making tights and stockings. Norun, or in other words, 'no run' was meant to imply the tights wouldn't ladder. Unfortunately this had to be vetoed as it wasn't actually true. But Norun had a certain simplistic ring to it and so was tweaked to Nuron ('Norun' backwards) but vetoed because it sounded like the nerve medicine Neuron, then tweaked again until Gladding landed on Nylon. It sounded good, and importantly didn't sound like anything else, but it meant absolutely nothing. Unusually, nylon is just an alternative name coined for polyamide and unlike many other names in common parlance, is not a trade name in itself.

Before moving on from nylon/polyamide, it is worth taking a brief diversion to consider its cousin polyimide, a material also developed by DuPont. This material is more than just a letter away from polyamide. It has emerged as an incredibly strong material in its own right, far and away stronger even in fibre form than nylon. It is best known by its trade name, Kevlar.

Here we reach a point where a massively important contribution to the world of polymers was made by a woman, this woman being Stephanie Kwolek, who also came across a new material quite by accident. Kwolek grew up in America, had a love of sewing and fabrics inherited from her mother, and nature from a father she sadly lost when very young. She also developed a passion for science and medicine. Facing a dilemma over her future career, she plumped for a career in medicine, but decided to start in chemistry.

Kwolek began in the research laboratories of DuPont, where she was tasked with investigating possibilities for finding a fibre with heightened capabilities under extreme conditions. She began experimenting with polyamides as her starting point. She happened across a method which formed these materials into a fibre with lined- up parallel lengths of molecular chains. Having the chains oriented all in one direction made for an exceptionally strong material. Even better, the fibre forming, or spinning, process was remarkably straightforward, and so Kevlar came into being. It is now used for so much more than body armour, finding its way into cabling, bicycle frames, golf-club shafts, automobile brakes, canoes and even the reinforced bodies of aircraft. And if you have ever watched the classroom trick of making a nylon rope appear from a beaker of chemicals, then you might be interested to know that it was Kwolek who developed this demonstration technique.

Polyimide has developed even further and, for those interested in space exploration, it may be worth noting how a form of polyimide has made itself indispensable here also. The launch of the James Webb telescope has been followed by many, after all this is the telescope that potentially can view the very origins of the universe. A key part of the telescope is the sun shield, tasked with deflecting the burning glare of direct sun rays. This shield was made of several wafer-thin layers of polyimide topped with a thin coat of aluminium foil.

The next polymer on our list, polyethylene (PE), continues on the trend of accidental discoveries, having manifested itself unexpectedly in Chester in 1933. Eric Fawcett and Reginald Gibson were two scientists working for ICI in the 1930s when they heated a mixture of ethylene and benzaldehyde under extreme pressure and realised a white waxy substance. Fawcett probed a little further and designated it as a polymer of ethylene. It had potential, but unfortunately ICI decided that the whole process getting to this material was just too fraught with safety hazards. Given the rudimentary equipment of the time, the entire processes was a disaster waiting to happen with a high risk of explosion. So the experiment into PE had to be shelved on safety grounds. Luckily by 1935, with better equipment, the research was allowed to continue. Now another team at ICI, using ethylene as a starting point, tried to replicate the outcome. They did, and out popped a lump of PE.

As it transpired this was a serendipitous conclusion that took a little while to replicate. The original vessel used had leaked, allowing oxygen to add itself to the mix, which as it turned out was pertinent to the process. Once this was realised and factored in, a method for producing what was to become known as Low Density Polyethylene (LDPE) was formed. It had taken the team at ICI five years in total to reliably replicate the initial experiment, but in 1938 they managed to unleash polyethylene onto the world on an industrial scale. Some may now weep at the unfurling of a material that was to transform itself into the plastic bags much condemned now, but at the time this was a truly remarkable material and came exactly at a time when it was most needed. The world was about to enter another world war, and it proved an effective insulating material for radar cables. PE was also instrumental in allowing the instalment of a telephone line between the Isle of Wight and mainland England. In fact, PE offered enormous assistance during the Battle of the Atlantic and arguably helped save many lives. Post-war uses were initially a little thin on the ground until during the 1950s the Hula Hoop was born, which really brought the material to a mass market.

While we're on the subject of polyethylene, it's worth looking at another variation of this family of materials, and that is its high density form, more simply known as HDPE. Have a look at some shampoo or plastic milk bottles and you may find this stamped on the base. In original form LDPE was formed by a *high*-pressure process. Conversely it was found that HDPE could be obtained by a *low*-pressure process. In 1953 a German professor Karl Ziegler discovered a method of packing ethylene molecules together in a more uniform fashion (the tighter the packing the higher the density), and he could do all of this using a lower pressure and temperature. The Italian chemist Giulio Natta aided in perfecting and simplifying the process. His process went on to create the related polypropylene (PP). So important was the work of both Messrs Ziegler and Natta that in 1963 they were jointly awarded the Nobel Prize for chemistry.

The next material to cross our path is really the prelude to another better-known polymer, but is added here as it recognises a contribution by a woman in the field of polymer science. The woman in question is a Minneapolis native called Patsy O'Connell Sherman. During her school years in the early 1950s, Patsy sat an aptitude test that advised her best career choice was to be a housewife. Digging her heels in, Patsy insisted she would be given the same test the boys sat, and this time the indicators suggested a career in dentistry or chemistry. She selected chemistry, took a 'temporary' job with 3M, where she remained for her entire career (retiring in 1992) and despite the protocol of the time, she both married and had children!

Early in her career Patsy was tasked with finding a rubber-like material that could withstand attack from aviation fuel, as it was destined to be deployed on jet fuel lines. While working on a compound she accidently spilled a little on her shoes. It was only after some time she realised that her shoes were grubby everywhere but the spot the chemicals had coated. Intrigued, Patsy investigated further, and in time the spilled polymer which was found to also repel water and oil became the perfluorochemicals forming the basis for PTFE, better known under the tradename of Teflon. Patsy's rubber-like fluoropolymer also found a use and was marketed by 3M under the trade name 'Scotchgard' where it became a familiar sight in shoe shops as a spray that would protect both leather and suede and even fabric. However in a postscript to this story, the environmental impact this material has on the planet is increasingly being recognised. It will not degrade naturally, and this fact coupled with the unpleasant chemicals that made up the fluoropolymer has led the producers, 3M, to phase out certain elements in the chemical composition of this product (they are known as PFCs).

From the 1950s (as mentioned earlier), there was an explosion in plastics production, as it was discovered that they could be used as a substitute material for all sorts of products: e.g. PVC medical drip bags instead of glass, polypropylene fuel tanks instead of metal, polyester fabric instead of silk, melamine picnic ware instead of delft and polyethylene chairs instead of wood. Even the newly patented Velcro (1955) was found to work better using nylon and polyester rather than the original cotton version.

By the 1960's those workhorses of the polymer industry had been discovered and commercial production begun. As we have seen some plastics had been chemically discovered surprisingly early but only found to be useful decades later, polycarbonate for instance, being a good example. Its earliest appearance comes in the test tubes of a German chemist in 1898 but not patented until 1953 and then commercially produced in 1958. It is probably from the 1950's onwards that we see a shift with plastics becoming mainstream materials in their own right.

There has of course been further development in new materials through intervening years but in the main it is those mainstream materials that have been tweaked and filled and blended to deliver a plastic that will give just the properties needed for an application. New developments now tend to come in the form of highly specialised materials able to serve very specific markets. So now we have PTFE, acetal and polyether sulphone amongst many other materials we would find it very hard to do without. The Covid pandemic really showed this to be the case with so much PPE wholly or partially reliant on plastic. In the early days plastic was new and used primarily as a substitute material. Plastics really took off in this respect as they were easy to mould, lightweight, colourful and cheap. Soon the adaptability of this material was noticed and new varieties, blends and filled grades were developed and these found their way into products unimagined before the advent of plastic. The development of nanomaterials is just one small example of the continued advances that have been made. Plastics are now all around us.

The genie is out or the bottle, but do we really want to stuff it back in? Don't forget early development of plastic came about to halt the march towards extinction of creatures we would be horrified to even think of losing. Elephants, whales and tortoises may not have been with us now had plastics not been developed when they were and the insatiable desire for their parts continued unabated.

We'll look again at the uses of plastics in different sectors and acknowledge their role in changing society. Before that it's worth throwing in a little science to explain the workings of plastics in a little more detail. All too often they are just pitched in together with one descriptive word – plastic, but there is so much more to them than that, with many nuances and facets that make them ideal for so many applications. Knowing how they are categorised and how they can be expected to behave under particular conditions will go a long way to aid in their understanding as a material.

3

CATEGORISING THESE MATERIALS – THE SCIENCY BIT

As already touched upon, there is an extensive range of plastics available nowadays. Some are more common than others. They can be broken down into broad categories, namely Commodity, Engineering and High Performance. Price and usage tend to determine which plastic goes into which category.

Of these the commodity plastics are those that will generally be more familiar. These are materials that don't have any inherently special mechanical properties and are relatively cheap to produce. They include polyethylene (PE), polypropylene (PP), polystyrene (PS) and PVC. These make the pots, tubs, bags, trays and bottles we are so familiar with today.

Engineering plastics are those that can be used in applications where they can utilise their superior mechanical properties. These are the plastics you will find in hip replacements, car parts, safety items, protective ware, glazing, bearings, gears and building products. They include the very versatile PET (aka polyethylene terephthalate) and also plastics such as nylon, acetal, polycarbonate and acrylic.

High-performance plastics are more specialised. They can withstand higher temperatures and tolerate difficult working conditions in both chemical and environmentally challenging applications (here we are talking about high-humidity wet, or otherwise hostile conditions). These are found under the car bonnet, in aviation, electrical and optical applications. They include the lesser-known polyether sulphone (PES), polyethererketone (PEEK), polyphenylene sulphide (PPS) and polytetrafluoroethylene (PTFE, better known by its trade name, Teflon).

We have already seen why plastics came to be derived from oil, but how can oil – and gas – make all these different types of plastics? Fractioning and cracking the oil gives us ethylene, and this in turn gives us the building block we need to produce other plastics. It might be worth mentioning that ethylene is a by-product of the fractioning process, which means that only a small amount – around 4 to 6% – of oil we extract becomes plastic.

So taking this ethylene, which we know to be lengths of molecules containing bundles of carbon and hydrogen, we can adapt and manipulate the molecules into something else entirely. Arranging these bundles (or molecules) into particular shapes and sizes will give different plastics. The conditions under which this manipulation occurs will also influence the type of plastic made.

For instance, production (otherwise known as polymerisation) of polyethylene under low pressure conditions yields High Density Polyethylene (HDPE). HDPE is used to make crates, bags, pipes, milk bottles. Ethylene put through a high pressure process conversely yields Low Density Polyethylene (LDPE). LDPE is used to make a whole host of products from carrier bags to packaging containers including squeezy bottles, carton liners, bin liners and cling-film.

HDPE tends to be less pliant, while LDPE tends to be more flexible. Of course the thickness of the plastic will also be influential, it is possible to have a HDPE carrier bag just the same as one of LDPE, but the former tends to be a little more robust. Now you can also find other densities of polyethylene from linear low (LLDPE) to medium (MDPE), each giving their own nuanced capabilities. LLDPE, being stronger, can be used in ever thinner gauges in packaging applications including the film lining on paperboard and MDPE, having better resistance to cracking, is used for piping.

Many other plastics are possible using different configurations of the basic carbon-hydrogen bundle, while others use a combination of additional chemicals all incorporated with our initial molecule of carbon and hydrogen. One analogy I like to use is that of building a wall with bricks. The way the basic brick is laid down can affect the end result. So laying the bricks in a certain pattern can give a herringbone pattern for instance, or indeed any number of other patterns such as stretcher bond, Flemish bond, stack bond, English bond, etc. Sometimes a builder can introduce other building materials like flint or stone, wood or daub. It all gets incorporated into the structure but yields different results. At its most basic it is a wall, but depending on the components added, it can be utilitarian, strong, good in certain environments, or just aesthetically pleasing.

Another method employed to create further varieties is blending. Blending can yield materials according to manufacturer's requirements in the same way that whiskeys or coffee for instance can be blended to taste. They can be blended in such a way as to give particular properties and make them adaptable to all sorts of situations. So PP and a rubberlike material known as EPDM can be blended to form a material that produces very effective car bumpers.

For all their abilities and adaptabilities, plastics and rubbers do have natural handicaps. For instance a number are sensitive to light, ozone, heat and so on. By incorporating any number of additives, otherwise sensitive plastics can be made useful in all sorts of circumstances. These additives can be as innoxious as the chalk fillers used to make lino (added to PVC) or talc (added to polypropylene) used to make car door panels, right through to waxes, pigments and oils. Some additives do contain chemicals known to be problematic, but these are tightly regulated and several are now banned (such as those containing lead).

Some basic plastics applications use no additives at all other than pigments, but in some cases we can see why additives are needed! Have you ever left plastic clothes pegs on the line only to find them snapping to bits in your hand? Pegs by and large are made from polypropylene. This is a great material for all sorts of reasons and a great many applications, but one of its biggest downfalls is its behaviour when exposed to UV light. The UV attacks the bonds holding the plastic molecules together and because polypropylene has a high degradation rate, the bonds break and the pegs snap. They are a cheap product, so manufacturers don't always feel the need to add to their costs and incorporate additives to stop them breaking. You can always go out and buy more pegs, which of course the manufacturer is very happy for you to do.

Environmental effects come in all sorts of forms, from the yellowing of PVC to the crazing of acrylic beakers. This is one of the reasons that additives are used. They can slow or stop some of these effects and extend life. Additives can be used in all sorts of other ways of course, whether to make a rigid plastic flexible (such as PVC), harden a plastic to make it useful in a greater number of applications (such as HDPE or PP), strengthen and toughen to add to a plastic's natural capabilities (such nylon and polycarbonate) and so much more.

So these are the origins and categories of plastics as we know them. Now we have quite an extensive range of formulated materials at our fingertips. Materials that undoubtable have proven very useful to mankind. Why is it then that we have such an adverse reaction when we hear of products being made of plastics? How is it that some plastic products can be esteemed and valued, but not necessarily thought of as plastic? Maybe we should next look at the terminology commonly used and cut through some of the jargon. We can also look at another way in which these plastics can be categorised, and this relates to their recycling capabilities.

4

THE LANGUAGE AROUND PLASTICS, FROM MEDIA TO MARKETERS

We have already looked at some of the terminology and descriptors around plastics and rubbers but more do crop up. Some terminology is used by the media to portray these materials in a specific way, usually derogatory, yet some is used to impart worth and desirability. But what does it all mean? How do we distinguish between them? Both the terms polymer and plastics have peppered the pages so far, so maybe this is a good starting point.

Polymers v. Plastics

The media have developed their own shorthand when informing us of plastic-related news. When an article sets out to demonise plastics, they are called just that – plastics, whereas when the material forms the basis of some new wonder stuff, a scientific breakthrough that is going to save lives or make the world a better place, it is often referred to as a polymer. So what's the difference?

We earlier discovered that the term 'polymer' was a catch all term referring to both plastic and rubber but what is a 'plastic' and what is a 'rubber'? In the early days it used to be quite clear-cut. A plastic referred to a material that was malleable while a rubber was a material that was stretchy and elastic. Rubbers would remain stretchy and elastic and fixed in their given shape during their lifespan. This was because they lightly crosslink on manufacture. They lean towards those thermoset materials that cannot be reprocessed. They weren't plastic (even though they were initially malleable), their elastic characteristics meant they were something else, and that was rubber. It was a term that has been around for a long time and so understood.

Over the years, as rubbers further developed, they moved away from their natural beginnings and became increasingly synthetic in form. Now some rubbers can also be referred to as elastomers. The original understanding of the term 'rubber' is no longer quite as distinct from 'plastics' as before. With ever more adaptable synthetic rubber being developed, the clear distinction between these materials has blurred. Now there are such things as thermoplastic elastomers. Essentially these are rubber like materials that can regain malleability when heated and reform, yet are also stretchy. Silicone is a good example of a hybrid material that can straddle both camps. It is an elastomer that can be formulated as a thermoplastic or a thermoset yet has rubberlike elastic qualities. Rather than leave these, and other, elastomers in a limbo of terminology, it became increasing common to just call the whole kit and caboodle, polymers. It really is a more descriptive term anyhow.

So the term 'polymer' can cover plastics, the in-between thermoplastic elastomers, and rubbers. It can even be used to define certain natural materials – and so for that matter can the term 'plastic'.

If plastic just means a malleable material, what does the word 'polymer' mean? The word can be broken down to its constituent parts – 'poly' and 'mer'. The 'poly' bit just means many, but what is a 'mer'? Basically it refers to a part, in this case a unit or group of atoms that, when repeated many times, make up the building blocks of plastics and rubbers, and of course of thermoplastic elastomers. Scientifically this group of atoms, all bonded together, is called a molecule.

Lots and lots of 'mers' all forming a very long chain can produce a very useful material. These long chains (or molecules) entangle and impede each other, making solid materials that can be shaped into useful products. So for instance four hydrogen atoms stuck onto two carbon atoms give the simple molecule ethylene (C_2H_4). Stick lots of these ethylene molecules together and you get a material that can make bags, bowls, milk bottles, trays and a whole host of other products.

Not so many 'mers' tied into short chains can also be useful. Short chains can slide across one another with relative ease. Short chain polymers can be found in products such as washable paint that easily be slapped across walls without too much effort from the painter. These are usually vinyl materials. Another common short chain polymer is PVA (polyvinyl acetate). This is commonly seen as a glue and in glue like products.

Now that we have looked more closely at the meaning of the word polymer, it might make the concept of thermoplastic and thermoset materials easier to understand. Particularly when it comes to figuring out why some polymers can be recycled and some can't. We need to go back to the molecular chains to find out why. An analogy might make help.

Let's visualise a molecule, our basic chemical unit, now think of it joining a conga line with other molecules. Our conga line is a molecular chain. Where the conga line is short it can easily slip around other conga lines provided they too are short (remember the glue and paint examples of short chain molecules). As the lines get longer, they can still move (once heated) but with a little more difficultly. Useful plastics are made from lines of molecules that have an optimal length where they don't slide too easily past each other but also aren't so long that they overly entangle and can't move at all (even when heated).

All right, we now have a number of conga lines of optimal length nicely warmed up, all congaing. They all head to a circular room and have to line up to enter through the door, once in they fill the room nice and tightly packed, it gets cool and they can't move any more. This parallels what happens to a plastic that is heated up, pushed into a mould and cooled down until solid. Now with a thermoplastic ('thermo' refers to heat), we can get these chains going again so they can become mobile. This means they need to be reheated. So off we go with the conga lines again, they are out of the circular room but now head into a square one. Some of the lines might get a little shorter as the conga can break when it squeezes in and out of the rooms, but they make it with mostly optimally long lines and some a little shorter, and a new shape is made.

So it is with thermoplastic materials. These are the ones that are most useful when it comes to recycling as they can be chopped down to little pieces, re-melted and reformed several times. Somewhere in the region of 85% of polymers produced fall into this category. Remember though, the chains do shorten a little each time the material is chopped down and passed into a mould, so there is a limit ultimately to how many times this can be done. It is usual to mix at least some unused virgin material in with the recycled when it is remoulded just to keep a higher level of optimally length molecules. Luckily all the commodity plastics and a lot of the engineering plastics are thermoplastic, so all these pots and trays and bags can in theory be recycled, as can all sorts of other products they are used for such as buckets and bins and bumpers. Why they aren't always recycled will be looked at later. That brings us on to the thermoset materials. Sticking with the useful conga analogy, we're back with our lines of molecules all sliding happily passed each other. Off they go herded into the first room but this time while they are there the temperature doesn't lower to fix their shape. This time think of a conga line of people were each line has a person at set intervals along its length who will put out an arm and grab someone from a neighbouring line. They won't let go. Now the lines or chains are firmly bound to each other and no matter how much shoving goes on, each individual conga line can't break away and move into a free state again. No matter how much heat is applied, their lines are fixed solid.

In plastics this firm hold between chains is called crosslinking. These links are caused by a chemical reaction, sometimes hurried along with the addition of heat, which sets the shape and it can't be reversed. This is why these plastics can withstand high heat without melting. Although they won't melt, each of these materials has a temperature limit beyond which it will just burn. So if you have an epoxy or melamine worktop you will know that you can place hot things on it without melting it, but if a pan say is really hot you can scorch these surfaces. Many of the high performance plastics mentioned earlier fall into this thermoset category.

Rubber sits between these two categories. It is lightly crosslinked. That is any given chain will only have a few links to neighbouring chains. This allows the rubber to stretch and pull away a little between chains but ultimately it will just snap back to its original position. As a chemical reaction has taken place and cross links exist between chains this does mean that the material can't be melted down and reformed. Unless of course it is a thermoplastic elastomers, then it has the best of both worlds.

So to recap, plastic is a polymer and a polymer is a term that can be used to refer to a plastic, rubber or thermoplastic elastomer. The next time a news article refers to a polymer being used for a noble cause consider, is it a natural occurring polymer or is it synthetic? If the latter, then there is evidence of the usefulness of this much maligned material.

Polymer isn't the only term used when a retailer doesn't want to say their goods are made of plastic. I'm sure many of you will have come across the adjective 'resin' in many settings. You can find this term popping up for all manner of goods from kitchen sinks to side-table ornaments. But what is this resin, and is it too a plastic?

Plastic vs. Resin

Marketers and retailers are very good at making goods sound up-market and imbued with high intrinsic value. Plastics conjure up the idea of cheapness, and why wouldn't they? This was a strong early selling point. The problem is that the notion of cheap sticks as does the synonym of them being valueless.

When a shop wants to sell a 'resin' ornament say, one they want their customers to display proudly in their homes, they aren't going to say, 'Here's a lovely item made of plastic.' They will say something like, 'This stylish resin product will sit well in any interior.' Have you ever wondered where this resin comes from? Subconsciously we probably associate resin with tree sap and so think of a natural material. This is not the case. These resinous ornaments are very much plastic. They aren't quite what we immediately think of as being a plastic though. After all there tends to be a weight to them and we associate plastics with lightness. They also tend to have a richer finish, looking like bronze or some other higher value material. But this is just a surface finish, as a scratch would show.

So what is this resin? For cast ornaments the resin tends to be epoxy (but can be acrylic or polyester). Epoxy falls into the thermoset category of polymers, that is to say it changes from a pourable material to one which has set solid by a chemical reaction, crosslinking molecular chains of the polymer together. This locks the shape into place, but does mean that it can't be melted down and reprocessed. This is one of the difficult polymers when it comes to recycling. Its thermoset characteristics are just what are needed when epoxy is used to bond fibres on an aircraft wing or boat hull (another of its many applications), but can be problematic when we get fed up of our ornaments and want a change of decor. There is limited recycling scope for these resinous ornaments.

Of course epoxy isn't the only polymer masquerading under the catch-all term of resin as hinted at already. Polyester, related to PET but this time in thermoset form, is also thrown into this category. Have you ever seen bathtubs or worktops being advertised as being made of 'cultured marble'? This isn't some fancy opera-loving marble, it is crushed rock, sometimes marble or limestone, sometimes quartz, sometimes a mix, bound together by polyester. Less cultured more plastic. Now this term has been around for a while, manufacturers are mixing it up again and calling it engineered stone. A greater variety of crushed stones can be used but the binder remains either polyester or epoxy.

Acrylic is another 'resin', used in these sorts of applications. But this is cast acrylic and another thermoset. It can look pretty in a bathroom, be coloured to match other units, but can scratch. Cast acrylic crops up frequently enough under the guise of resin also.

We'll maintain our interest in epoxy for the moment though. A little-advertised fact about epoxy is that it tends to be stuffed full of Bis-Phenol A, otherwise abbreviated to BPA. Here is another area where marketers can be misleading with their labels. BPA-free labels are slapped onto products with gusto. All sorts of products that never had, and never will need to have, a whiff of BPA in their manufacture. Yet BPA is used to make epoxy and epoxy is widely found in these 'resin' ornaments, so why are there not warning labels on these products if this is such a dangerous chemical? This question will be looked at next.

5

TO BPA OR NOT TO BPA

What is BPA and why are manufacturers so keen on telling us their product is BPA-free? Why should we be so worried about BPA, and are there other chemicals we need to be equally concerned about? What is it even doing in plastic, and for that matter what types of plastic is it in? So many questions, so let's try get to some answers. We'll start with BPA itself.

Bis-Phenol A (BPA) and its Safety

Bis-Phenol A, more commonly seen abbreviated as BPA, is a phenol-based chemical that has known effects on the human endocrine system, but it continues to prove useful in the manufacture of a variety of items we see daily. We'll look at these as we further examine this chemical but they do include certain inks, thermal papers used for some till receipts, adhesives, additives within certain grades of PVC, some flame retarders and even brake-fluids. More on this later.

Negative effects are known to be connected with BPA, but under very particular conditions, not least of which is connected with dosage, in other words, the amount that would need to be ingested over a given period of time. An in-depth and objective investigation was instigated by the US Food and Drug Administration (FDA) under the auspices of their National Toxicology Programme which was designed to identify potential health effects across a wide range of exposures to BPA.

The report is on-going (at the time of writing), but observations released to date state that, although the scientists conducting the investigation have observed changes in the pituitary and reproductive tissues of the rats used in the study, these occur only at the highest BPA doses administered. These levels of exposure are many times higher than the levels people can typically expect to receive in normal conditions. The report goes on to acknowledge that at lower doses some health effects were recorded; however, in this instance it was not apparent, and could not be said with certainty, that the changes were related to BPA exposure. Initial analysis led to a statement by the Deputy Commissioner for Foods and Veterinary Medicine at the FDA, who said: 'our initial review supports our determination that currently authorized uses of BPA continue to be safe for consumers'.[4]

This is by no means the only research conducted on this chemical, as many others have come before, but this one is the most recent. Other reviews into BPA have been carried out by a number of governmental food and drink agencies including those of Switzerland, Canada and New Zealand, to name but three. (Of course the European Food Safety Authority has also reviewed the use of BPA in plastics, specifically that used in food packaging.)

This chemical has been around for a long time and therefore has undergone rigorous testing; testing that for fifty years has shown it to be safe in the areas to which it is put to use. Yet testing continues nonetheless. There is no doubt that in certain quantities agents within BPA are harmful and therefore its usage is strictly controlled. A Tolerable Daily Intake (TDI) had been established by the European Food Safety Authority and set at 4 micrograms per kilo of bodyweight per day. Even so, BPA had been found in levels well below this TDI so it was unlikely that the average consumer would be faced with harmful levels.

If a product needing the attributes conferred by the chemical but does not use BPA – a substance, as we have seen, with a long test record – then what is being used instead? BPA is used for a very specific purpose and often can't be just left out entirely without recourse to a substitute; certain can linings for example may not be possible with a BPA alternative. Tweaking and substituting is all well and good, provided we know just what is being used as the substitute chemical, and scientists just may not have the same level of certainty of the safety of these possible alternatives. A label can quite truthfully say BPA-free, but quietly ignores the fact that BPF or BPS may have been used in its stead.

So where BPS or BPF and so on (the list allows for a fair bit of the alphabet to be swapped for the A) are concerned, are they really any better than BPA? Who knows, as the research isn't there to the same degree as with BPA, but as BPS is bisphenol S and BPF is bisphenol, are we really any better off, when it is the bisphenol bit people are mostly concerned about? Somewhat bizarrely BPF was found to occur naturally in mild mustard,[5] yet it too is a chemical that can affect oestrogen levels in humans. Do we need labels on mild mustard telling us to swap to English mustard? In fact, BPF also occurs naturally in several of the seeds and rhizomes used in traditional Chinese medicine. Studies[6] as recent as 2019 have advised that these naturally sourced herbal products ought be considered a potential source of endocrine-active bisphenolic compounds having the potential to contribute to overall exposure levels.

An element of caution is needed when reading articles warding us off certain products. We need to consider who has commissioned the research or article, or even who has written the report, and to question what response they are rooting for. What are the driving factors for the authors and is there a hidden agenda? Don't just believe the first article you see if it agrees with your point of view. Ask why they are putting forward a particular narrative. If the data is presented in an unbiased manner, giving all the details, it is likely to be more trustworthy than a report which actively seeks to highlight the negatives, or indeed the positives, without putting the data in context. And look for references to back up statements. If an article or website sets out on a diatribe but does not back up the claims made, then it may just be a personal opinion not based of any science. It is very important to get a true picture.

Another note of caution needs to be inserted regarding BPA. Labels occasionally appear on many types of plastics that would not be likely to contain BPA in the first place. They crop up on supermarket shelves, occasionally stuck on lunch boxes or refillable water bottles. I have seen them on polypropylene (PP) products, yet PP is one of those plastics made from molecules of carbon and hydrogen and doesn't need a whiff of BPA in its production. Certainly, water bottles can be made from polycarbonate (PC) – these are the clear, robust bottles – and PC for now relies on BPA for its manufacture (more on this in a moment), but opaque, slightly squishy bottles are likely to be made from PP. These labels are confusing and misleading. The manufacturers and retailers are *not* demonstrating great concern for our health (otherwise they would stick labels onto jars of mild mustard!), they are simply trying to drive sales by appealing to our caution where health is concerned. So where *can* we find BPA in manufactured goods, and does it only appear in plastic products?

Polycarbonate and epoxy are two major plastics necessitating BPA in their manufacture, so let's look at these materials next.

CHAPTER 5

Polycarbonate (PC)

Polycarbonate (PC) is a fantastic material used in a variety of products, from baby bottles to safety glasses and from bus-shelter panels to riot shields – not to mention such out-of-the world applications as astronaut visors. It may also now be seen used as screens in shops navigating their way through the aftermath of the pandemic. It is useful in all these applications because it is very clear and very tough and has a terrific resistance to impact. But what about all that BPA? Should we be worried that it is contained in these products?

Well, not really. Certainly PC is largely BPA-based, but we need to look at what happens to it when it's made. Remember what we said about polymers being created by getting bundles of atoms to join together to make long-chain molecules. This is called polymerisation. Well, if we look more closely at how these bundles or molecules join up, we would find that actually there are a couple of ways they can go about it. One method just involves adding one molecule to the next (think of the conga line). This rather simply is called addition polymerisation.

There is another method which is called condensation polymerisation. Here groups of chemicals are placed together; they react forming a different type of chemical and jettison the left over molecules. This is called condensation polymerisation because in order for the groups to join together they need to make room on their molecular chains. They do so by shedding certain other atoms which bundle up to form secondary substances that can be as innocuous as water but can also be as noxious as carbonyl chloride. All this happens at the raw-material manufacturers. Once the polymer has been created it goes off to the product manufacturer, who moulds it into any number of products.

Polycarbonate is made by condensation polymerisation and uses BPA in its initial step. (BPA itself is formed through a condensation reaction, where phenol and other chemicals such as acetone are reacted together and give off water as a side product.) For polycarbonate production, BPA is reacted in solution where a number of different chemical combinations are possible. BPA is the common factor, whatever the other chemicals chosen. This means a number of secondary substances can be produced, some unpleasant and others more acceptable. (It is worth saying here that the industry is aware that some very unsavoury chemicals can be used in this process and continue their investigations into more neutral alternatives.)

Once this chemical soup has been made and the necessary reactions undergone, the secondary substances are washed out, leaving the polycarbonate. Now that the connections are made, that's it: it's polycarbonate, just like if you beat an egg into milk and apply heat to make it an omelette. You can't pick out the egg from the milk.

Likewise, left to itself, once BPA has made polycarbonate it's not inclined to separate back out again. We do need to check if any stray unpolymerised bits of BPA have been caught up with the polymerised polycarbonate (similar to an undercooked omelette with some runny bits trapped inside). This is where scientific research comes into play.

The Danish Environmental Protection Agency in 2015 conducted a very thorough investigation into just this topic – that is the migration of BPA from polycarbonate and its propensity or otherwise to leech out of PC. They looked at PC intended for food, medical and toy use, both in ambient conditions and immersed in water at a high temperature. First they looked at the residual levels of BPA present in PC (that's the unpolymerised material trapped within the PC) across all their samples and found it to lie between 5 to 80 ppm (parts per million) with a DVD giving the higher value of 80ppm, while all food-related PC, including baby bottles, recorded no higher a value than 13ppm.

Focusing on the food-grade PC, the study looked at migration that occurred when immersed in water at 100°C (so, for instance, if a cup of tea was made in a PC cup, the only difference is that they kept the temperature at 100°C for an hour). The migration levels found were well below the level then recommended by food agencies. A number of the baby bottles tested recorded levels of migration below detectable levels. Next they subjected the PC products to hydrolysis, which effectively broke down the PC. The plastic was put into harsh environmental conditions and left at elevated temperatures (90°C) for two weeks, with migration levels checked daily. Even under these extreme conditions migration levels remained low, with levels remaining for the most part well under 2.5ppm. To clarify, these migration levels refer to chemical movement exiting the material as opposed residual chemical trapped within the material. Not all trapped material will migrate which is why the levels quoted are different for each.

That doesn't mean investigators ought to close the book completely or get complacent. Testing is always useful, especially as new means of testing are developed. However, and this is important, although it is unlikely that stray BPA will migrate out of a polycarbonate in its everyday usage to any significant degree, problems can be exacerbated where the material is heated. Heating can cause chain mobility.

Remember the conga line of molecules waiting in their room until they were heated. Once at a certain temperature they start to become mobile. This does not mean they need to be heated enough to flow (melt), but just enough for them to be mobile. This can allow enough movement, or wiggle room, for stray unbound molecules (those of BPA for instance) to come to the surface, where they can then pass into other substances.

So where a PC container is holding a liquid or food, it theoretically could become incorporated into these. This is why you are advised not to microwave certain containers if they contain foodstuff. Microwaving in particular can exacerbate movement due to the way it heats. It is also best if broken or crazed PC products are discarded if they are used for food contact. In general it is good practice not to warm up food or liquid in anything but those containers designed for heating.

Indeed, as mentioned above, testing is useful and does continue. In the very latest update, in April 2023 the European Food Safety Agency (EFSA) published a re-evaluation of BPA's safety. They drastically reduced tolerable daily intake (TDI) of BPA by a whopping 20,000, taking it from 4 millionths of a gram per kg of body weight to 0.2 billionths of a gram per kg. It is interesting to note that the UK's Food Standards Agency provided an update in June 2023 stating that 'the level of BPA detected in food to date in the UK is not currently considered to be harmful'. They have not so far altered their TDI recommendations. Maybe it's a case of watch this space.

Epoxy

Moving away from PC, several references have already been made to BPA being used to produce epoxy, so now we can turn the spotlight on this material. Epoxy is known chemically as Poly(Bisphenol A-co-epichlorohydrin), which gives a clear hint as to its chemical make-up, so let's start by looking at how it is manufactured.

It quite simply starts off as the combination of BPA with another chemical called epichlorohydrin. These two chemicals are mixed together in varying proportions, depending on the desired end result, and this initial mixture of the two chemicals gives a monomer – the starting point before it becomes a polymer.

Many will have come across epoxy as a two-part system. You start with one liquid, add another, swirl it around and then apply it quickly before it sets. Well, industrially it is not too much different, although here it can be produced as a solid or liquid. This BPA-enriched monomer now needs a catalyst in order to induce it to set. This inducement to setting is called curing, and the reactive agent can be any one of a number of chemical substances, occasionally just another phenol-based chemical.

Once all the mixtures are completed and epoxy formed, what happens to any leftover BPA? There is always a possibility that this residual unreacted BPA can leech out, as we saw with PC, but it is unlikely to occur in any great quantity, as the vast majority of the chemical does react in the monomer creation process. Scientific studies have been conducted, although not in any great number, and those that there are tend to concentrate on the use of epoxy in can linings (or on PC). None appear to present any findings showing that BPA leeches to any significant or harmful levels. Nevertheless strict regulations control the use of epoxy in certain applications with proscribed levels of cure that must be met. It is this curing of the epoxy that locks the chemicals in place.

Going back to the monomer, where BPA has been mixed with its companion chemical, at this point the manufacturer may wash out stray BPA. This would be the point where the monomer is prepared for the next step in the operation.

Maybe we should now consider where we use epoxy. Its many and diverse uses may surprise. Epoxy is used to manufacture so much more than the 'resin' ornaments mentioned in an earlier chapter. It is found in an array of domestic and industrial settings, where it is used to great effect. Domestically, it can be found, for example, throughout the kitchen, especially if you happen to have cast worktops and sinks.

Here epoxy can be used in various ways, sometimes as a top layer sealing in the materials below, and sometimes as a binder for smaller particles of marble or stone. Very handy when making these stylish surfaces more affordable. The alternative of course would be to hew large slabs of stone or marble into worktops and sinks. These are heavy, cumbersome, prone to cracking and need to be removed from the natural environment, and you can achieve the same effect from cast epoxy alternatives using waste or smaller particles of the stone.

But why are worktops not flagged up as a major health concern? After all, they take up a large surface area and are in regular contact with food, and can even be used as a cutting surface. Maybe the BPA these materials contain really doesn't migrate in health-debilitating quantities or maybe the comparatively short contact time means this doesn't flag as an issue? Whichever may be the case this is one instance where very strict guidelines are in place. These dictate that where epoxy is used in such applications, it is of a suitable food grade and cured to an extent whereby the surface cannot allow migration of sizable quantities of BPA.

At the time of writing, I have been unable to find academic papers researching the migration of BPA in applications such as kitchen countertops. This may be due to those strict regulations and guidance as imposed by governmental agencies such as the FDA (covering the United States) and the European Food Safety Authority (its European equivalent). These regulations also cover DIY applications of epoxy. For instance marble makes a beautiful countertop but has the drawback of being porous. Spilt red wine will tell its tale if not mopped up quickly. Tomato juices too leave a lingering reminder. Epoxy forms a protective layer and can be applied retrospectively. It must however be applied properly, be of the correct grade and cured fully. Those grades available for such applications at the local DIY store will have been formulated to conform to such applications and will give instructions on how to achieve the maximum cure. On a side note it might also be worth mentioning another DIY use of epoxy and that is in the craft sector where poured epoxy can be shaped to make rings, broaches, bangles and any number of decorative items. Again if properly cured the BPA content locks within the epoxy.

There is another use for epoxy that puts it in even closer contact with foodstuff, and that is its use as a liner for 'tin' cans and inside the lids of glass bottles and jars. It can even be put to use coating the machinery on food production lines to prevent them from excessive wear or corrosion from the foodstuff they process, transport or store. It can also be used to line certain pipes conveying potable water.

Although some cans are lined with other materials, such as tin, epoxy is used to line cans for quite a large range of foods, particularly in the UK, as it provides a very good barrier between the metal of the can and the food it contains, so preventing oxidation and the ingress of bacteria. This is very useful, as some foods can be quite corrosive in nature. According to the Epoxy Resin Committee of the Association of Plastics Manufacturers in Europe,[7] even the unlikeliest foodstuffs can cause problems. The humble pea, if left to its own devices, has a tendency to blacken the inside of cans, tomato can make them red, while pickles soused in salt and vinegar can just weaken the can entirely. Even soft drinks are pretty corrosive and could cause the can to rust over a short period of time. Putting a tarnished penny, or even a tooth, in a glass of Coke for a day will soon demonstrate the corrosive powers of this beverage. The epoxy lining, which is incredibly thin, prevents this reaction from taking place.

A greater potential for BPA migration comes when temperatures are elevated during processing; this indeed is the case for all cans lined with epoxy. Although to put this into context, the levels of BPA found to migrate are still measured in parts per billion and well below the levels flagged by the Food Standards Agency. Once again, I would suggest it is good practice to avoid heating containers, including cans.

Epoxy can be seen in many industries, ranging from colossal wind turbine blades used to create green energy to protective coatings found on anything from the humble washing machine to a gleaming car body, from the easy clean floor of a shopping centre to a ship's deck and hull. It can also provide structural components for aircraft and automobiles (more on this in a later chapter), golf club and tennis racket shafts, fishing rods, skis and a host of other sporting equipment, as well as in the electrical sector (again this will be looked at in a later chapter). None give pause for concern as regard their BPA content.

Other uses for BPA

As we have seen, BPA is used in the manufacture of some plastics, but not nearly as many as we are led to believe. And while various forms of bisphenol are used in a number of applications, but what about bisphenol-A specifically. Can it be used in common applications other than plastic and what about our exposure in these instances?

BPA can be used in the production of thermal paper, which is used on all manner of receipts found in a broad range of outlets. It is fast and efficient and doesn't require messy and expensive ink cartridges. Turning standard paper into thermal paper involves the surface addition of chemical layers, principally BPA. In thermal paper the BPA is not embedded within the material, as with certain plastics, but rather it sits on the surface, where it can be transferred far more easily to the skin. Studies have shown that BPA can be found on thermal paper in concentrations up to 42,600ppm and anything between 8 to 17ppm (parts per million) can be transferred to the hand.[8]

This is why the EU decided to restrict its use in thermal paper from 2020. At time of writing this is not a total ban, as the European Chemical Agency is allowing time for its phase out.[9] Because of the chemical mix found on these thermal papers it is not recommended that they be recycled. Indeed the Dutch government conducted an in-depth study[10] on BPA giving specific emphasis to its behaviour in the aquatic environment. Findings showed that the process of de-inking waste paper when recycling resulted in comparatively substantial quantities of BPA being released into waste water. The study recorded emissions in water to be 202 kg/year and 403 kg/year for PC and epoxy respectively yet recorded a figure of 151,600 kg/year for recycled thermal paper.

Migration of other chemicals from plastics[11]

We looked at the migration of BPA from various sources, so here is a good spot to say a little more on the topic of migration of other chemicals from other plastics. Worries over the migration of chemicals from polymers are nothing new. During the 1980s concern was expressed over PVC, and more recently over polystyrene (PS) also.

PVC was very effectively used as cling film, but there were worries that cling film in direct contact with food which then was microwaved would cause additives within the PVC to migrate into the food. PVC was also widely used in tubs, bottles and other containers.

PVC is inherently a stiff rigid material, so to make it useful for the packaging industry in particular a raft of additives were needed. These included stabilisers, plasticisers (to make it flexible) and various other ingredients, including BPA in some instances. These were all worthwhile at the time because of the advantages PVC gave as a packaging material, not least because it didn't allow gases to pass in or out, which kept things fresh.

With PVC the concerns centred on the oils used as plasticisers. The problem was particularly relevant where fatty foods were concerned. PVC is now not commonly found in food applications; polyethylene tends to be used as cling film these days. It is not as effective as PVC, but satisfies the caution demanded by consumers.

Polystyrene has also undergone some changes in application. Here the concern centred on free benzene rings. Benzene is a concern as it is a carcinogenic, and these benzene rings are found in the monomer. Remember the monomer refers to one styrene unit before it is joined to its neighbours during the polymerisation process. As we have seen with PC, some monomer may be present in the end plastic, where it hasn't managed to join up with a neighbour. This would not cause a problem under normal conditions, but when PS is heated, particularly in a microwave, this could pose a problem.

Polystyrene is used to make tubs and trays, although these are now increasingly made now from PET. It can still be found in the likes of yoghurt pots and chocolate box trays. It is also used in an expanded form for takeaway containers and hot drinks cups. There it provides good insulation, keeping the meal or drink warm. When the contents are kept in the containers and heated in a microwave the monomers within the plastic can start moving. They can then rise to the surface, where they potentially then can cross into whatever is touching that surface, such as food or liquid. Certain elements within some foods can pick up monomer more than others. Fats, greases and tomato are particularly troublesome in this respect; this is why advice is always given not to reheat anything in these containers.

A final point on this topic worth mentioning concerns a report undertaken by the UK Food Standards Agency which included an overview[12] of the potential migration of contaminants that can occur in packaging manufactured from natural sources, including paper and board, particularly where these have been recycled. The problems arising from recycled paper include migration and contamination from inks used for printing. These can contain unsavoury heavy metals. Residual pesticides can also be problematic, as can natural toxins such as those developing from moulds (fungus). However, as this 2019 study acknowledges, only limited research on migration has been undertaken for bio-based materials.

It does highlight the fact that issues can arise from the most unexpected sources, not just plastics. We look on synthetic materials and expect them to cause us problems, and in doing so overlook the materials that are supposed to be better for us and the planet.

6

WHEN IS A BIOPLASTIC NOT A BIOPLASTIC?

Bioplastics are a topic invoking much confusion and yet little is done to enlighten the public. As a result, these materials may be causing more harm than good in environmental terms. The first developed plastics could be termed 'bioplastics', but that did not mean they could biodegrade once their usefulness expired. Modern plastics have the very same issues. It is worth noting that bioplastics can be broken down into three separate categories: those that are truly bio-based and biodegradable, those that are bio-based and are not biodegradable, and those that are known as drop-ins – where traditional fossil-based plastics are hybridised with incorporated elements from a bio source – and are not biodegradable. To add to the confusion, it is also possible to have non-bioplastics – i.e. fossil-fuel-based – that are biodegradable.

The problem, as far as public perception is concerned, hinges on the fact that few bioplastics actually biodegrade. The terms bioplastic and biodegradable are not synonymous, despite what we are led to believe. Sticking the prefix 'bio' onto a word can sometimes convey a sense of reassurance, when in fact the exact opposite may be true.

Lately, enticing eco-emblazoned labels have been slapped on products hoping to attract environmentally conscious consumers. Shoppers selecting these goods, despite their best intentions, may in fact be making matters worse. Bombarded with messages telling us to make the right choices, we are given scant information on the true meaning of these labels. So how can we possibly know which product is environmentally the better option? It is confusing and one that many unintentionally get wrong.

Throw into the mix the pithy symbols that now proliferate, particularly in the packaging sector. Unfortunately they often do little to help either. All are eye-catching and appealing but not necessarily informative, and so this might make a useful starting point in unpicking issues surrounding the claims of green or bio products.

The particularly useless symbols are those that signify a potential to biodegrade without saying how. A prime example is the brown looped seedling symbol, meaning the product has the potential to degrade once composted. Most likely this does not mean in the average compost bin at the end of the garden, and it certainly doesn't mean it will naturally rot down if carelessly discarded.

Some symbols are utterly misleading, intending to convey a reassuring message but in fact say very little. A good example is the round 'green dot' symbol made up of two entwined green arrows stamped on various forms of packaging. At first glance this gives the impression of recyclability, and the fact that both arrows are shades of green also implies biodegradability. In fact this European symbol just means that the product manufacturer has made some sort of financial contribution to the recycling or recovery of the packaging. It doesn't mean that it *will* be recovered or even that it *can* be recycled, or be in any way biodegradable.

The 'mobius' symbol – i.e. the one that's roughly triangular in shape with three consecutive looped green arrows – also conveys recyclability. Here at least the product does have the potential to be recycled, but it does not indicate that it actually will be post use. Sometimes several of these mobius symbols appear on packaging, one for each material used. So a carton may have a plastic film overwrap, a cardboard shell and a plastic window. Each individually may be recyclable; they may all even be accepted for recycling. However correct recycling would depend on the consumer deconstructing the entire package and placing each component part in the correct receptacle. Not everyone feels the compunction to be so thorough with their refuse.

The 'widely recycled' symbol depicted by a heart on the end of a circular loop again signifies potential but is not a statement of fact. Some councils may take it, others may not. Even where it is taken, it may still be rejected if certain requirements aren't met, for example if there are food residues or other contaminants present. Changing messaging adds to the confusion. Initially advice required all lids and caps be removed before recycling. Now they need to be left on plastic bottles but removed from glass bottles and jars. Indeed there is now further development in some caps for fizzy drinks where they are designed to flip rather than unscrew entirely, a little like the aluminium ring pulls that used to pull away from cans but now push inwards.

And if you think this is a minefield, it gets even more confusing when we look at some of the wording and post-use claims bandied about. Let's begin by deciphering the commonly seen terms 'bioplastic', 'biodegradable' and 'compostable'.

Bioplastic

'Bioplastic' is one of the most overused and meaningless terms around at the moment. There has been a noticeable increase in the amount of products, primarily packaging, now labelled as being bioplastic. Strictly speaking this label is not inaccurate but it can certainly be misleading. As we have seen above, bioplastic is bio-based; this does not mean it is biodegradable. Products proudly and prominently display this stamp of apparent green credentials, but what is a bioplastic if it's not biodegradable?

A bioplastic is one that has been produced from a bio material such as corn, bagasse (the fibrous remains from sugarcane or sorghum), all manner of starches and a number of other mostly plant-based materials. It can even be made from algae. Its origins may be completely green, but not necessarily the end product.

Many bioplastics have their plant-based foundations manipulated, jigged about and altered to an extent that they end up as a plastic, a proper plastic equivalent in nearly all respects to its oil-based cousin. Remember that plastics at their most basic level are bundles of chemical elements, many created just from atoms of carbon and hydrogen. Where these chemical elements come from, whether oil or plant, and how they get bundled into plastic-producing molecules really doesn't matter. They have the same properties, same functions, same potential applications and the same problems when it comes to end-of-life issues – that is to say, they *will not* biodegrade. To further complicate matters, some plastics can also lay claim to being 'partially bio-based'. This is an additional layer of complication that just means the resultant plastic was created from mixed sources, renewable and fossil-based. There in all likelihood is no indication of the ratios or proportions of each constituent, so the bio element can be very scant indeed.

Problems are exacerbated further when a consumer disposes of a bioplastic product in the incorrect recycling stream. Unfortunately most bioplastics cannot go into the compost bin. These products for the most part need to go into the plastics recycling bin, because this is what they are – a plastic. Conversely, problems can also arise where a plastic is genuinely formulated to degrade but looks and feels like a standard non-degradable plastic. If it is unwittingly thrown into a recycling bin for plastics it can sully the whole waste stream. These two forms of plastic are immiscible.

To give an example of the misconceptions and potential problems surrounding bioplastics we can look at PEF, a.k.a. polyethylene furanoate. PEF is a type of plastic known as a polyester, and so is very similar in nature and properties to PET. It can be derived from agricultural crops or even from bio waste (such as for instance chicory roots or indeed from many forms of fructose). It was discovered some years ago but shelved for a number of decades before coming to the fore again in answer to the increased demand for biopolymers. It was previously shelved because of technical barriers and processing difficulties in its production. High energies are needed to convert the fructose sugar to PEF, not to mention the need for corrosive acids and lengthy time-consuming processes, all of which are environmentally impactful. So all in all, this bio alternative is neither green nor economical.

PEF was recently 'rediscovered' when it was found to be useful in similar applications to PET such as pots, tubs and trays. In some aspects it even offered enhanced properties, certainly in packaging applications. Overall it is so chemically similar to PET that it will happily enter the same waste stream – in other words both materials can be recycled together.

There is no doubt that those looking for a bio-based alternative for their packaging, would find PEF a useful material. Its potential sparked much ongoing research to alleviate its recognised drawbacks. In a few years we may well see PEF products sitting comfortably on the shelves with PET. But whatever 'bio' sticker ends up on these products, we must remember, PEF *does not* biodegrade no matter what newspaper articles or press releases say to the contrary. I have already seen respected broadsheets declare PEF to be the next degradable wonder material. It is not, it's important to reiterate: once the raw material has been converted into PEF it has become a plastic just like PET.

Now we need to consider our priorities when it comes to plastics. Is the drive to bio-based plastics all-consuming and the ultimate objective? Consider the following example: we want to make a safety visor and have a traditional plastic just right for the job. There is a bio alternative; in this case PET or PEF can be used. Now do we want that visor to come from a plant-derived plastic no matter what it costs environmentally? The issues of expended energies and crop appropriation are often overlooked.

Energy is also required to create and form oil-based plastics, although here the raw material is derived from the by-product of petrol production. But don't forget energy is needed to isolate the building blocks of any bioplastic from its crop source. Additional energy is then needed in both cases to take the building blocks and arrange them to form a plastic. Whatever the drive towards green energy, much is still produced from incinerating fossil fuel. Do we on the one hand wish to save fossil fuels by not using them for plastic production while incinerating the same fuel to create energy for converting crops to plastic?

What about the raw materials need to produce bioplastic? Where are all the plants coming from? We are again being given mixed messages. Space is scarce. Rainforests are being cut down and replaced with crops sought by those in the West. We are being asked to increase our awareness of the source for a whole range of crops which enter our lives in many guises.

Of course, many of the plants suitable for bioplastic conversion can be grown locally, but not all by any means. Even where we eliminate the transport costs of the more exotic crops, modern-day farming on a large scale requires water, fertilisers, pesticides, herbicides, heavy machinery to reap and space to grow. The use of GM crops has been touted as a means to achieve better yields, but if this is the case and the resultant material degrades back into the soil, will this cause contamination? It is a dilemma. Mention has already been made of the role oil plays in creating certain substances such as fertilisers and pesticides, not to mention the CO_2 and even methane emissions created when breaking down this organic matter. Even where biomass comes from waste products – how much waste is produced using standard farming procedures? Does it fulfil all requirements if we are to source only renewables to make plastics in the future? We can't fix one problem by creating another potentially worse problem.

There is no easy answer to all of this. It boils down to looking at all the available information and doing what seems right on balance. There is no quick fix that will solve all the world's problems, particularly if we continue to crave all that we currently consume. Now with that in mind let's move on to biodegradability. We have encountered the term many times, but is it as clear-cut as we like to think?

Biodegradability

Think biodegradation and chances are you think of the natural breaking down of a material. Indeed when we talk of biodegradation, we are considering the natural breakdown of a material through attack by microorganisms. The process may be natural, but it is certainly not straightforward, with many external factors needing consideration. In other words, a plastic may be biodegradable, but with what caveats?

For instance, do we want a discarded piece of litter to degrade in air if it's thrown on the side of the road or tossed from a car? What if it is buried? Again there is a difference between burial in compost or soil, and even how deeply it is buried, not to mention the soil type (is it sandy or clay for instance?). And now the question of microorganisms comes into play, as well as any incidental effects on soil-dwelling creatures such as worms. After all, a material breaking down will produce some sort of matter. Whether this matter will be harmful to soil-dwelling creatures if ingested is uncertain. Remember, a bioplastic doesn't grow in this form, it needs to be extracted and manipulated using all manner of processes and chemicals.

Even the temperature and oxygen content of the soils matters, as will be seen in more depth when we consider compost. And what happens if it ends up in water? Again there is a distinction between sea water and fresh. Elements within the water itself will also have an effect on degradation as do oxygen levels and temperature.

A plastic with biodegradable credentials may very well degrade in soil, but not at all in the sea. Another might quite happily disappear to nothing in a ditch but not a river. This is where those often cited standards come in to confuse shopper and retailer alike. Standards are there to regulate and give consistency in their targeted area. They provide uniformity, reliability and reassurance. In other words, if a product is marked with a standard that says that product will degrade, then that is exactly what the product should be manufactured to do.

As a result, a manufacturer usually has several standards from which to choose. Some standards are very narrow and others don't conform from one global region to another. Certain degradation types have no standards available. Biodegradable plastics are not straightforward and so no one standard can cover all eventualities.

To illustrate the point, there exists a new European standard covering biodegradability in soil. This is the pithily entitled EN 17033, a.k.a. the OK Biodegradable Soil Label for Mulch Covers. As inferred by its subtitle, it relates only to agricultural biodegradable mulch films. These are the covers used in farming for crops; they impede weed growth and help retain soil moisture. This is the first international standard looking specifically at biodegradability in soil.[13] It specifies the requirements for biodegradable mulch films designed to be ploughed back into the soil without creating any adverse impact on the environment.

There are still issues to be addressed in this particular standard, not least of which is the rather generic approach to soil classification, since the performance of a degradable plastic will alter depending on the soil type. The standards also specifies a target of 90% degradation across two years, but what about the residual 10%. And what if there is a build-up of un-degraded material in the soil because of course crops tend to be planted annually. Much still needs to be addressed, and this is just for one very specific, very targeted area of application.

Whatever the shortcomings of providing regulations for soil, at least there is a starting point. No reliable standard exists for biodegradability in water, either fresh or salt. And no standards are available at all (at the time of writing) to cover biodegradability in air, i.e. when a discarded item is left to break down in the atmosphere. Composting, although attracting its own issues, does have attendant regulations and that will be covered in the next section.

The UK government, wishing to clarify the situation, have published a consultation document looking at just this issue. This document defines biodegradable plastics as being those that can break down into component parts such as water, biomass or even gasses (carbon dioxide or methane),[14] and acknowledges that external environmental factors come into play during the degradation process.

The document recognises the potential for an imbalance in the ecosystem following biodegradation, but the degree of potential toxicity arising from biodegradation is just not known. Any standard developed would have to give an end point. Would a bag be counted as having degraded if visible shards are present? Or must any remnants diminish to microscopic levels indistinguishable from surrounding material? Does it transform into the natural substances expected such as methane or biomass and indeed CO_2, or are there chemical traces of the plastic remaining? If so might this not just lead to the creation of additional microplastics? Defra go as far as saying that currently they don't have enough evidence favouring the widespread uptake of biodegradable plastics. Their evidence doesn't indicate whether it would 'increase resource efficiency, reduce waste, or tackle plastic pollution'.

Microplastic production aside, what about the gases produced on degrading. Many are considered to be greenhouse gases. Really it is swings and roundabouts when trying to find the better option. According to an article by Columbia University,[15] bioplastics can create less than half the greenhouse gas emissions than their oil-based counterparts since they can balance the carbon emitted with that absorbed when growing. However, depending on the manner in which they subsequently degrade and breakdown, they can emit methane. Unfortunately methane is markedly worse and more damaging as a greenhouse gas.

There are other often forgotten aspects that just add further complications to this issue. One particularly may be minor but does need considering and that is the question of inks. Many plastics, again predominantly in the packaging arena, have had various information printed either directly on the product or on a plastic shrink film which is placed on the product. It the product and shrink wrap overwrap degrade, what happens to the layer of ink?

There are several regulations in place covering diverse geo-graphical areas (for instance the EU, USA and Switzerland) that control the constituent parts of inks used in food packaging, particularly where the ink could potentially come into contact with the foodstuff. The regulations ban the use of ink ingredients such as those that are toxic or contain heavy metals. This is fine if these restrictions cover biodegradable food packaging, but what about other biodegradable plastics? If they too carry printed information, what happens when the host plastic degrades? What if the product degrades at a quicker pace than the ink, might this leave potentially harmful residues? Don't forget, the much cited BPA can also be used in the manufacture of these inks. It all needs consideration if we truly want to cause the least possible harm to the environment.

Indeed it is not only the environment that needs consideration in these cases. We need also consider the impact on ourselves. Several UK universities are researching what has been termed bio-based food contact materials and their results are shared with the Food Standards Agency.[16] Data from these and other sources were analysed to provide science-based evidence of the potential risks and other unintended consequences in replacing traditional fossil-based food-contact plastic with that of bio-based materials, including biodegradable bioplastics.

The study aimed in part at ascertaining the presence of agents and contaminants including allergens, bio-toxins and nanomaterials. They acknowledged the potential risk of contaminants, which could include heavy metals, pesticides, veterinary medicines and natural toxins such as those occurring from fungal infestations. Basically the concerns voiced do not only look at potential soil contamination but also at potential food contamination during use. Ultimately there is a need for further research in this area to clarify the situation, but the highlighted risks are interesting and again often overlooked.

Another worthy consideration concerns the impact on waste streams should a truly biodegradable material end up in the wrong recycling box. Many biodegradable products aim to ape mainstream plastics and, to all intents and purposes, can look just the same. This only serves to add to consumer confusion. Slinging everything into the same waste stream could end up corrupting a whole consignment of waste. Likewise, thought is needed in order to convey this to the general public and indeed to those collecting our recycling material also otherwise the potential for unprocessable waste would be enormous.

This still leaves us with the problem of trying to distinguish between biodegradable and compostable. And even between domestic and industrial composting – there really is a world of difference between these, so let's see if we can untangle this next.

Compostable

So, how can we distinguish between the terms 'biodegradable' and 'compostable'? It can be argued that particular environment effects are required to initiate composting, whereas a truly biodegradable product would break down naturally. A case could also be made that under these circumstances a compostable material would break down faster than one decomposing naturally. The caveat being that composting necessitates very particular conditions being met, and these normally refer to temperature, oxygen and moisture levels. Moreover, compost ought have the ability to be added to the land without going through any further degradation or emitting odours. The UK organisation WRAP (Waste and Resources Action Programme) makes a useful distinction saying that not all biodegradable plastic is compostable, but all compostable plastic is biodegradable.[17]

Laying these basic differences aside, what of composting itself? Here we find another minefield in terms of terminology. If you think being compostable means chucking waste in the garden compost bin where multiple microbes and bacteria can do their work, unfortunately that is just not the case with plastics.

Two forms of composting are available, both quite distinct from each other, and unfortunately this is really not being made clear to the general public. Before looking at the differences, let's first clarify what outcomes are on the wish list for a compostable plastic. At a very basic level, a compostable plastic is one that can be added to soil after a managed breakdown, where it provides some sort of benefit to the soil such as a fertiliser or soil conditioner. Basically we want something to break down in such a way that it can provide a boon and be indistinguishable from that soil.

How this can be achieved involves either one of two very different techniques, home composting or industrial composting. Home composting usually involves the regulation council compost bin. There is a lid keeping contents dark and a certain associated warmth depending on the time of year and the waste being added. Microbe content also can vary greatly according to the material added. Even the technique of turning compost heaps makes a difference to the rate of decomposition, as do moisture and aerobic conditions. This is all a country mile from the highly regulated industrial process.

Industrial composting uses high heats to break down a material, much higher than can be found in a home compost bin, typically needing a sustained 60°C or above. It also needs a combination of higher humidity and oxygen, again much greater than would naturally be found in garden composting. This together with aerobic and nitrogen regulation offer a highly controlled biological decomposition. Quantities processed are far greater, but the size of the waste added can maintain prescribed dimensions through grinding and shredding processes prior to composting. This all aids in speeding the process.

Interestingly there is a compliance standard for composting (EN 13432). Seeing this on a package confers a level of reassurance. In addition, the compost symbol (as touched on in the start of this chapter) is increasingly appearing on certain items. Do either of these stamps mean the product requires industrially composting or can it be done domestically, or are both options open? This is yet more confusion for the consumer. Industrial composting can work on a far greater range of compost-compliant products, far more than anything ever achievable at the bottom of the garden. The symbol does not always make this clear. As for the standard, it only applies to industrial techniques.

Surprisingly, there is no standard for home composing. However, there are some individual national standards and factsheets available covering home composting, such as that released by European Bioplastics.[18] This seeks to compare available information as offered by a variety of European countries. They identify only two European regulatory bodies that offer a certification programme. Although broader standards for products suitable for home composting are not currently available, this is an area under investigation.

Now we have established the meaning behind these increasingly prevalent terms what about the even more ambiguous 'green' products put on our shelves? So-called green or environmentally friendly materials are being promoted as valid alternatives to more common traditional plastic materials. Some are listed as being biodegradable or compostable, although not all by any means. It would be interesting at this point to look at some a little closer and determine their true green credentials.

Just how good are green materials?

In an interesting study by Plymouth University,[19] researchers left a variety of carrier bags, acquired from a number of outlets, in different environments. The bags included a number of conventional polyethylene, compostable and biodegradable examples, which were assessed at regular intervals. They found that bags left in the open degraded into fragments over time, whatever their make-up. Compostable bags buried in soil remained intact after three years but disintegrated in the sea after three months. Other bags buried in soil or submerged in the sea and retrieved after three years were extant and robust enough to carry a full load of shopping. This study demonstrates the complexities involved when trying to understand the true green credentials, or otherwise, of so-called alternatives.

This study aside, let's look at a few examples of materials being lauded for their supposed green credentials, marketed on the strength of their natural plant origin. Bamboo, PLA and cellophane are interesting starting points. Here we'll take another look at the potential for cellulose to be touted as a green material.

Bamboo

Bamboo became very popular and is very on trend. So called 'eco-cups' and all manner of containers are made from the stuff. Even Cabinet Ministers like to display their green leanings by clasping these cups during photo ops. Such products may originate from bamboo fibre, but have you noticed how these products feel to the touch and how weighty they are in the hand?

They do not have the texture or lightness of a product manufactured from a fibrous grass. There is a very good reason for this, and it's all due to the manufacturing methods used. The process includes a vital yet surprising addition for an apparently eco product.

The production starts with small fibres of bamboo, fibres that are powdered to break down to minute sizes. Powdered bamboo isn't going to be much use as a container, so now these tiny fibres need to be bound together again. Melamine does the job nicely; it not only binds the fibres but gives it that degree of heat resistance necessary if the finished product is going to be used for hot beverages. This heat resistance is also very useful in making any ensuing product dishwasher compliant.

But hold on a minute – isn't melamine a plastic? Certainly it is. Not only is it a plastic but it is a thermoset plastic as mentioned earlier. Remember that one? This is where the plastic is formed through a one-way chemical reaction that crosslinks the material in such a way that it cannot be melted down and reformed. This is the type of plastic which is typically the most problematic for recyclers, i.e. it cannot be recycled conventionally. This is before we even begin to consider the common accessories of a silicone lid and sleeve. Silicone, of course, is another polymer.

Mixing melamine with organic fillers (in this case bamboo), effectively traps the fibres within the matrix material. This makes it difficult to find an effective recycling route for these products. One consumer group[20] suggests the best disposal method is to burn them, which would of course release unpleasant chemicals, including formaldehyde. Try burying the products and you'll end up with no degradability, as the organic bamboo fibres are embedded within the plastic. And they do have to be completely embedded otherwise they would wick out any liquid contained in the cup's interior. This would make for one damp cup.

As a multi-use plastic cup, it's reminiscent of 1950s melamine picnic ware, although it could do without the addition of bamboo, which really is doing little to help environmentally. It is acting as little more than a filler.

In fact such is the current concern over this form of 'bamboo' cup, governmental authorities from the Benelux countries (Belgium, the Netherlands and Luxembourg) issued a joint letter in 2021 stating that 'business operators will have to withdraw "bamboo-melamine" materials and articles from the EU Market immediately'.[21] This is because of concerns surrounding the use of bamboo as a filler and migration of chemicals associated with the melamine. Their concern highlighted the unauthorised use of bamboo in plastic cups and expressed a desire that consumers be notified of their findings. They particularly sought action and testing for such items used by babies and small children. All in all, it's probably best to avoid these products particularly if the environment is a concern.

PLA (Polylactic Acid)

Another material commonly pushed as being both environmentally friendly and biodegradable is PLA, a.k.a. polylactic acid. It is found in a range of products including the aforementioned cups and many packaging applications, as it makes a useful substitute for polyethylene or polypropylene. Although increasingly utilised, PLA is more costly to produce to traditional plastics and offers reduced mechanical and physical properties.

It is tricky to state categorically which camp – good or bad – this material sits within. It really does depend on a number of circumstances. Let's start though by identifying the source of this material. Here we are on better ground in environmental terms. The lactic acid needed for PLA is indeed plant-based, coming usually from maize or sugar beet. The lactic acid is derived from glucose, and while these crops in the past predominated in supplying the raw material needed, it can now come from a variety of additional sources. These include tapioca (it does have its uses outside school puddings!), sugar cane or wheat amongst others. Researchers are looking to push the source of raw materials further to include stock such as straw, bagasse wood chips or even, according to Belgian research, cheese waste.[22] The resultant sugars are then fermented to generate lactic acid, which is in turn converted to lactide. This is purified then polymerised to give PLA.

The production steps taken during this process will ultimately influence the resultant PLA in terms of its ability to degrade. When at the lactide stage a certain molecular configuration will create a material that is easily compostable. This form of PLA goes on to form short-shelf-life products such as packaging or other disposable items. A different lactide configuration will produce a form of PLA not so easy to degrade. This is used for products that require greater durability. These nuances are not so apparent to the consumer. Both forms of PLA can be labelled as PLA, but the outcomes for both can vary quite significantly when it comes to their degradation capabilities. This is made yet more complex when blends of both are used.

This brings us to the next important point concerning this material. PLA is very picky in choosing an environment in which it will degrade. It can degrade almost completely if given the right conditions, and the right conditions entails a trip to an industrial composter. Popping PLA in the garden compost bin will do little to encourage degradability. Likewise, landfill is not going to aid degradation if a PLA-based product ends up there. (Worse still of course would be a trip to a recycling plant in a plastics recycling bin, where it could contaminate the whole batch.)

As the required conditions of heat, oxygen and organic substrate are usually not met, any PLA product disposed of in landfill or even garden composters will remain intact for who knows how long. It needs an industrially composter, which is not always made clear on packaging, in part because those adding the labels are not always cognisant of the nuanced differences in terms.

Another common use of PLA is in the filaments used for 3D printing. Here we have another concern over and above the question of degradation, and that is the additional compounds present in these products. Whereas PLA moulded directly into products such as packaging is pretty straightforward, being for the most part just PLA, 3D printer filaments tend to incorporate far more additional elements in order to aid the printing function. These additives are not always as environmentally friendly as hoped. These filaments likely contain VOCs (volatile organic compounds), which can be released when heated, for example, and can be toxic.

Cellophane and other cellulose-based plastics

Cellophane is an interesting material. It has been around a long time (incidentally, cellophane is technically a trade name, albeit one that has come into common parlance). On the face of it, it should be completely biodegradable. But it really depends on how much tweaking has gone into making it into a useful material.

It starts well enough, with its key ingredient, cellulose, coming from natural sources, whether wood or plant. To usefully convert the raw material, it requires a rather complex procedure. In fact, the complexity of the process was one of the reasons it was dropped in favour of the easier PET and other such plastics some years ago.

To convert cellulose into cellophane, a pulpy starting point needs to be attained. This is achieved by breaking fibres down in an alkaline bath. Next up is an aging process where it is left for a few days then treated with carbon disulphide. This yields viscose. To get the viscose back to cellulose it needs another bath, this time in dilute sulphuric acid and sodium sulphate. Now it is cellulose once more, and better still, in a workable form. From here it can be made into clear film, but it needs yet another bath to both bleach the film and remove the sulphur.

So far so good, as far as biodegradable potential goes (even if it does entail drenching in some pretty nasty chemicals along the way). After all, this is still cellulose, just in film form. The problem comes when demands are made on this basic material. If it needs to be impermeable to vapour or heat-sealable (e.g. as in envelope windows) or made suitable for certain automated processes, then coatings are required. Where these coatings are also biodegradable all is well. However, this is not necessarily the case. For instance the far-from-degradable polyethylene tends to be the coating of choice in many applications. This can't easily be seen and it does mean that a potentially degradable material is rendered problematic. Problems worsen when the material is still called cellophane (after all, the plastic layer is so thin) and expectations are assumed in terms of biodegradability.

Cellophane pops up in all sorts of film applications. One of the most easily identified is the twist-type sweet wrappers on boiled sweets. Cellophane worked beautifully as it had the ability to twist and stay twisted. A lot of other films spring back open. Try it. It also has that satisfying squeak when untwisted! It's also been used for tapes (think rolls of sticky tape, particularly Sello-tape) although now many use polypropylene as a base. Cellophane can also be found in membranes, bakery bags and coatings for glass and paper. These applications tend to require plastic-coated cellophanes. To throw an additional spanner in the works, often the easier and cheaper to produce polypropylene (PP) is used in these application. PP is naturally quite opaque in appearance but pulling the film, in other words orientating it, bi-axially (in two directions) gives a clear film that behaves in a very similar way to cellophane.

Of course, industry is always striving to innovate and work continues on finding biodegradable coatings that will perform as well as those used now. But it is probably well worth reflecting on the path taken to manufacture cellophane. It was superseded some years ago by other plastic materials that were cheaper, easier to produce and used a less toxic cocktail of chemicals during production.

The fact that cellophane's route to production is also energy-intensive doesn't help either. Why try to save oil resources by not making so much oil-based plastic, only to use more of the stuff in driving power stations to generate the energy needed to create the likes of cellophane? Remember only 4% to 6% of the oil produced goes into making plastic, the majority of the rest is incinerated in one form or another. Would it in fact be kinder on the planet to make sure we effectively recycle all plastics, *including* film?

Cellophane is probably the better-known of the cellulose-based plastics, but there are others. We've already seen that cellophane is biodegradable, but what of the others, can these retain full biodegradability yet be environmentally kinder in manufacture?

The cellulose nitrate we met in the first chapters still has its uses. Although not as widespread, and now available in grades that are not as combustible as once encountered, it can be found in lacquers and varnishes for nails or playing cards. It can still be found in ping-pong balls and certain photographic films. Its inherent explosive nature is put to good use in magic tricks where a sudden blaze of light is required and achieved via nitrocellulose flash papers. In terms of degradability this is less straightforward. Cellulose nitrate is a material that needs a particular environment in which it can degrade; it is not likely to do so naturally. Although early film is known to degrade, formulations now mean the material is less likely to do so.

Another cellulose-based material is cellulose acetate. This is also based on a natural material but tweaked to make useful which can involve a variety of additives. It is produced by modifying cellulose with acetic acids. These acids can be derived wholly from a biomass source, but industrially only a proportion tends to come from this source.

Although there are several uses for this material, it is probably more commonly seen in textiles and as the filters in cigarette butts. It can also be used to produce spectacle frames, combs, paints, fibres and filtration membranes. It is another material that is governed by its route to manufacture and can be produced as a degradable material capable of breaking down under a number of circumstances (in water, soil or compost) in a relatively short period of time. However, it can also be produced to be longer-lived and can be problematic when it comes to its degradability, as can be seen with discarded cigarette butts. At present you are probably more likely to encounter the slow-to-degrade variety, as research and industrial process have not quite caught up with each other in manufacturing a fully biodegradable and commercial cellulose acetate.

Other possibilities

Fortunately, this is an area attracting quite a bit of research. Consumer pressure demands an increased range of available biodegradable plastics. The more commonly available have been mentioned, but there is also continued research into investigating other biodegradable plastics capable of being viable alternatives to existing fossil-based plastics.

Polybutylene succinate (PBS) is one such material. It has actually been around since the onset of plastic materials, but as it did not demonstrate the potential noticed in other synthetic materials, it was shelved. Now it is attracting renewed interest. PBS can be derived from sugary plants such as cassava and sugar cane, but interestingly from synthetic sources also. Indeed, part of the process to producing PBS commonly uses petrochemical feedstock.

Once converted, it can retain its ability to degrade, although again the extent of this is dictated by the way in which it is processed. On the whole it does have the potential to biodegrade or compost. It can be found in agricultural films but also in teabags (more on this later) and the plastic lining of disposable paper cups. The problem is that polypropylene, for instance, can also be found in these applications and it is not always clear where this is the case. PBS has possibilities and is increasingly being investigated as a viable 'green' alternative but also has certain production issues it needs to overcome before meeting its full commercial potential.

Despite this, it is probably a material to watch out for in terms of its green credentials and capabilities.

Overall, it is worth pointing out that, for packaging particularly, more and more degradable alternatives are becoming available. Some are properly biodegradable, others compostable, either at home or industrially. One factor tends to be omitted when such packaging is been touted as better for the planet, and that is that they tend to advertise the constituent parts we want mostly to hear about, i.e. the ones from natural sources. In fact, such packaging is likely to comprise a number of elements from natural and oil derivatives to unite and give optimum properties.

So a compostable bag will do exactly what it says, but it may actually be a compound of PLA, starch and PBAT (Polybutyrate adipate terephthalate). Some of these elements bulk and some bind, and while PLA and starch are bio-based, PBAT currently is not (and therefore this packaging is not wholly a bioplastic). PBAT is excellent in terms of biodegrading, but it is oil-derived. It is a synthetic material. In terms of PR it might not be what the public wants to hear, but it does make these compostable bags viable and it does properly biodegrade, leaving behind no tangible microplastics.

In fact, blending is a useful method in aiding the degradability or compostability of a material. PLA can't be home-composted but can be made to do so if mixed with polycaprolactone (PCL), another oil-based biodegradable plastic. These blends can be used to make, for instance, compostable cutlery. Interestingly PCL is used by hobbyists because of its ability to be easily heat-softened and kneaded into shape and subsequently fixed when cool.

On that note, a final mention will go to polyvinyl alcohol (PVOH/PVA), not to be confused with polyvinyl acetate, also abbreviated to PVA. This is also a synthetic biodegradable polymer – in other words oil-based. Polyvinyl alcohol is water-soluble and used in a variety of areas including as coatings for paper, to strengthen fibres and as a blend for other materials. It can also be found as the plump little envelope that surrounds individually portioned detergent.

As we have seen, when it comes to properly biodegradable polymers, oil-based plastics are prominent in the pantheon of materials currently available to us.

Uses of plastics

Over the last few chapters we have established the meaning of the term polymer and looked at certain attendant issues. We have also explored some of the other statements used in their context and hopefully clarified the more commonly held misconceptions. Next we'll look at areas of use and no doubt will encounter more controversy along the way. Maybe some reflection on our own actions and consumption is badly needed, but let's see as we progress through just some of the many uses of polymer materials.

The four major uses for plastics are packaging, construction, electronic and electric, and transport. These break down[23] more or less as packaging (40%), the construction industry (19%), automotive (10%) and electrical/electronic (6%). We'll look at each of these starting with perhaps the most controversial, packaging.

7

PACKAGING

Time to talk packaging, the bête noir of the plastics industry. As previously mentioned, packaging accounts for something like 40% of plastic output and is by far its largest user. Once seen as the solution to many problems, it has had a massive fall from grace these last few years. So let's try untangle the issue and see where it all went wrong, at least as far as public opinion goes, and why it still dominates the market to such an enormous extent.

Plastic packaging was greatly appreciated in the early days. A shopping trip a few decades ago could involve toting large glass liquid-bearing bottles homewards, not fun. If you weren't relying on a milkman, carting home quantities of milk in bottles with no greater capacity than one pint soon built up the arm muscles. The rise of supermarkets meant an increase range of available goods, and without plastic this meant cumbersome alternatives, particularly for liquids. Putting our liquid needs into plastic bottles made them so much easier to carry and transport.

Glass is heavy and bulky and needs transporting several times – from factory to bottler to shop to home. Transport via lorry takes more fuel, not to mention additional lorries, a double whammy resulting from the weight and bulk of glass. A given quantity of plastic bottles can sit snuggly together on a pallet, but glass requires a thicker walled section to keep its integrity, meaning fewer can be placed on a pallet of equal proportions.

The size and capacity of the bottle is also compromised using glass. While it may be possible to produce a glass bottle holding, say, two litres, the weight of the bottle together with its contents would present handling difficulties to the average consumer, not to mention those in the transport industry.

Aside from the obviously heavy glass, this principle applies to all manner of packaging materials, including paper. To illustrate the point, let's just think of a shop taking delivery of carrier bags. The retailer can choose plastic bags or paper. Paper is more acceptable now than plastic, but is this really helping the environment? Plastic is lighter and can be made into bags that are far thinner than their paper counterparts. So not only can more can be packed into a box or pallet, this pallet is also lighter and as a result requires less fuel for transport.

This becomes apparent if we look at the comparative thicknesses of paper and plastic bags. A typical UK packaging company[24] listing the sizes of their various products advertise a standard paper carrier bag of 0.09mm thickness. They also list comparative vest type carrier bags (these are the standard bags found at the till) which have a thickness of 14 microns or in other words, 0.014mm. A simple division will show that the paper bags are around seven times bulkier, which is why they would take up more space when transported.

So for every one lorry full of plastic bags, seven lorries would be needed to transport the same quantity of paper bags. This example uses a robust plastic carrier bag as an illustration, but functional bags can go as thin as 9 microns (or 0.009mm). These thinner bags are typically found in corner shops or on market stalls. Conversely, a well-known clothes retailer has switched to heavier-duty paper bags for their customers. It of course follows that these are bulkier still, and are no more immune to rain, or, indeed, splitting, than their thinner counterparts. Of course, an option of shopping with reusable bags remains – more will be said on this later, but, be warned, the evidence from in-depth studies may surprise.

Generally the comparative weightiness of glass, paper and plastic can be subjective. Comparing densities would tell a more accurate story, as it takes account of both mass and volume. This really does matter when it comes to transportation as it does factor into fuel requirements. So how do the materials compare when it comes to density?

There are a number of plastic types, each with their own density, but generally the polyolefins found in most packaging have a density of roughly 1 g/cm3 (this is pretty much the same density as water). Paper is a little higher, coming in around 1.2 g/cm3 (again this depends on paper type, but here we are considering standard brown Kraft paper). Aluminium, as used in cans, is recognised as a light metal, but even so it has a density of around 2.7 g/cm^3 while glass has a density around the 2.5 g/cm^3 mark, again depending on type. Steel used for instance in tinplate cans comes in highest at 7.85 g/cm3. Now factor in the wall thickness needs of these materials in packaging.

Plastic, paper, steel and aluminium generally form relatively thin-walled packaging, with plastic marginally thinner than the given alternatives. Of these, metal cans generally are the thicker, although adding in corrugated sections serves to strengthen cans, allowing them to keep wall thickness at a minimum. Glass, however, tends to be a lot thicker. Anything from bottles of pricier water to roll-on deodorant requires sizeable lumps of glass for their production.

This gives a rough idea of the transportational needs of goods packed in a range of materials. Higher-density cans of, say, dog food would consume more fuel when driven to the shops than the equivalent quantity packed in plastic pouches. Chunky glass jars of mixed herbs will adversely affect fuel demands compared with the same herbs packed in plastic pop-lid containers or sachets. This is before we even start to consider the energy needed to manufacture glass, paper and, of course, metal.

Plastic packaging has become useful far beyond the realms of supermarket goods. Supermarkets undoubtedly do leap to mind when we think packaging, with food packaging coming out top, but it is worth acknowledging wider areas of application. For instance medically the advantages of using PVC blood bags instead of glass became clear in the 1950s and really became widespread in the 1970s. The importance of keeping sterile all manner of medical goods from gauzes and bandages to dental and surgical equipment is well known. This is most effectively achieved with plastic packaging. The now abundant surgical masks come in sealed plastic packaging. Months of Covid-19 news broadcasts, and perhaps even personal experience, have made us increasingly aware of those elongated cotton-bud-type saliva swabs being snapped from packaging and, once used, repackaged in plastic vials and sealed with plastic screw lids ready for virus testing.

Agriculture uses large quantities of plastics too. It accounts for 3.4% of overall plastics usage, but it is worth mentioning here as so much of it relates to packaging. Miles of plastic film is used agriculturally, much of it swathing the land to incubate seedlings or act as polytunnels. Yet more is used to package great rolls of straw or cover silage. Such use of this plastic enables seasons to be extended and allows early crops of various produce to hit the shops. Overwrapping straw or hay (packaging in its largest form!) allows it to remain in the fields until needed with a reduced risk of rotting. But it must be acknowledged that great quantities of waste are being generated. Not just here but also in the bags of packaged feed and fertilisers used for cattle and in the fields, plant trays, bale twine and even netting. What is done to recover and recycle this waste?

Actually there are now mechanisms in place to deal with agricultural waste, waste that had been, and still is to a degree, problematic because of soil contamination. Previously farmers tended to just burn their plastic waste, and the practice wasn't banned in Scotland for example until as late as 2019. There is now a national collection system in place in the UK to recover much of this plastic, with the aim of recycling where possible.

Of course plastic packaging can be seen across the board, boxing toys and overwrapping reams of paper (which seems somewhat counterintuitive). It is both a welcome necessity and a nuisance, depending on your viewpoint and the extent to which it is adopted. It might be useful to remember how things were before plastic came to dominate packaging.

Running barefoot across a beach in my childhood meant vigilance, always checking what was underfoot. Beaches and popular picnic areas often came with discarded bottles, broken or otherwise, empty tins and drifting sheets of waxed paper. Litter unfortunately is not a new phenomenon. It's still there, still an eyesore, still needs to be tackled, but it must be acknowledged that it is not the hazard it once was, undoubtedly from a human perspective.

Certainly wildlife doesn't see it that way, but then litter was always a hazard to wildlife whatever form the material took. Glass may eventually erode, but not until considerable time elapses. In the interim it can remain problematic for whatever it encounters. The same goes for many alternatives, including aluminium and, worse, rusting steel. Coir, hemp or cotton net or string bags, once widespread, may also be useful natural materials that eventually degrade, but do not do so overnight. While they remain extant they provide just the same hazard to wildlife as their polypropylene or nylon replacements. Even waxed paper takes time to break down and is not a natural food source for whatever creature consumes it. All these 'traditional' materials are unpleasant and a potential menace for wildlife. It is carelessness, laziness and litter that causes problems ecologically. All litter. The longevity of plastics litter is by no means exonerated. It exacerbates the situation immensely particularly where bioplastics are carelessly jettisoned with the mistaken notion that they will degrade.

Perhaps surprisingly, it's not just plastic that has a problem degrading or takes time to break down. The notion of packaging taking centuries to break down is not modern. A walk along the Thames foreshore in central London is very revealing. Firstly that there is surprisingly little plastic waste washing up given the population density, and secondly that previous generations of Londoners had no compunction in discarding their unwanted packaging. A walker on the shore at the Southbank for instance, right in the heart of London, will crunch their way across mounds of red debris. Yet the indigenous geology of London consists of grey and buff chalk, clay, Quaternary flints and gravel. A closer inspection reveals that the predominant red comes from sherds of broken amphora and pots, the packaging of their day. These have been littering the foreshore for as long as two millennia, back to the days when the Romans marched through London. It may be earthenware, but it's still there and is litter all the same.

After this small historical detour, let's head back to the supermarket.

Supermarket packaging

Plastic packaging, for all sorts of reasons, superseded other choice materials and perhaps became a victim of its own success. Barely a supermarket aisle can be traversed without coming across this ubiquitous material. It is used to package, to protect and to preserve goods for as long as possible. This abundance of packaging has not gone without notice. There is an increasingly loud call to reduce the amount used, but why did we need so much in the first place? To answer this we need to focus on the basic purposes of packaging, which can be summarised as to preserve, protect and entice. Let's examine each in turn.

Preserve

This is a primary function of packaging. There is no doubt that, used correctly, plastic packaging can preserve food and so extend its shelf life. According to a United Nations report, in the region of one third of the food designated for human consumption produced globally is lost or wasted. That equates to something like 1.3 billion tonnes. Considering just fruit and veg, the foodstuff we would normally find packaged in supermarkets, the statistics are worse, with up to 50% being lost.[25]

Eliminating or, at the very least, reducing food waste is unquestionably a good thing. It may not seem obvious just looking at, say, a cucumber swathed in plastic, but this artificial sheath is extending its life and keeping it fresh. No one wants to buy a droopy cucumber, and unfortunately they do have a tendency to droop quite quickly post-harvesting. Many high-water content vegetables do. Our constant demand for perishable goods such as cucumbers out of season doesn't help either. It would be pointless shipping in pallets of cucumbers from warmer climes if they drooped before even getting to the shelves.

This is of course only one example. Many fruits, vegetables and jointed or cut meat have a very limited shelf life depending on how prone they are to attacks from the atmosphere. Some fruits, such as bananas, give off ethylene (interestingly the same base as much plastic packaging) as they ripen. This process creates sugars, which attract bacteria, leading to spoilage. Left in normal atmospheric conditions, the process can continue unabated, but changing these conditions forces a slowing of the ripening process.

Eliminating oxygen (as is done on a localised scale by sheathing cucumbers) and upping levels of preserving gases such as CO_2 will slow the ripening process. For this reason certain packaging for some fruits, flowers and meat are pumped with CO_2 before being sealed. This is partly the reason why some bagged salad products for instance seems to be more air than leaves.

Meat is interesting. It has an extended shelf life when it is packed in what is known as Modified Atmosphere Packaging (MAP). MAP can seal in CO_2, but can also additionally retain a combination of gases including oxygen (O_2) and nitrogen (N_2). These gases are the same as those found in the environment but are used in different ratios than found naturally. Using gases the packaging can control biochemical and enzyme reactions, reduce moisture loss and inhibit the growth of microorganisms, all important in keeping meat as fresh as possible. Gas levels can also be tweaked to make the product seem more appealing. High O_2 levels for instance gives cut meat an appealing bright-red colour, as does carbon monoxide (CO), making it look fresh and appetising. After all, who wants to buy grey meat, whatever it says on the Best Before sticker?

Retaining all manner of fresh produce in the atmosphere most suited to keeping it fit for consumption, moisture-free and, of course, eye-catching, is achieved through its packaging. This calls for levels of sophistication not previously encountered in packaging. A simple polyethylene sheath as found on our friend the cucumber may not stop all ingress or egress. Different plastics have different properties. Sophisticated packaging must take the best elements from across a number of materials. Some plastics are good for blocking targeted gas ingress; others are good at keeping beneficial gases in. Therefore it is not unusual to have film packaging that actually comprises several extremely thin layers of different plastic types bonded together with each acting as a targeted barrier keeping the produce fresh.

This multilayered packaging can comprise anything from four to seven layers of film (although more are certainly possible), all just microns thick. It can turn up on more than just meat packaging. It can be found in the liners of cereal boxes, crisp packets, juice cartons and paperboard liners. Of course, this does not make recycling any easier even where film is accepted. To make matters worse, as far as recycling is concerned, aluminium can also be added to the layers to give a metallised film. Beloved of crisp and chocolate bar manufacturers, metallised film is fortunately not widespread in the food market. It is however quite common in the gift market, with foiled wrapping 'paper' and helium balloons widespread.

On a more positive note, raw material manufacturers are abundantly aware of the increased call towards sustainability and continue to develop materials that can truly biodegrade post-use. There are degradable materials out there, but the goal is to format a material that can confer all the attributes of existing packaging. It also needs to do all this within a price bracket that will encourage farmers or wholesalers to adopt any new packaging. In other words, it can't add too high a premium to the packaged product, otherwise the cost will transfer to the consumer and the goods risk remaining unsold.

Protect

Some packaging protects delicate produce from harm during transportation. Here rigid protective walls of plastic encompass soft or vulnerable goods from getting overly knocked about. This is a good thing and helps eliminate yet more waste. What may not be so obvious is utilising packaging to stop us as consumers from being too fussy.

CHAPTER 7

We all do it, given the chance – pick over a pile of fruit or veg to select the most succulent. Multiply this pickiness by the many customers in the grocery aisle on a given day, each picking over the stock on display, seeking out the ripest, choicest option, ignoring those with marks or bruises or those that are too ripe, too big or too small. At the end of the day the supermarket is left with a sad heap of rejected fruit or veg. How much better for the supermarkets to prepack a selection of fruit or veg for us so we don't even have to think about our choice.

From the supermarkets perspective, pre-packaged fruit and veg are convenient on several levels. It is easy to stack neat sets of, say, six apples per pack into the pallets or boxes that come from the farmer or warehouses with everything sitting nice and snug in its bag. Individual pieces of fruit and veg are restricted from rolling around where they are liable to damage each other and each can sit in its own controlled environment. Once in the supermarket, it is so much easier to stack these pallets onto the shelves and from there, for the customer to grab one from the shelf and drop into a basket. This all has a knock-on effect on price, of course. Goods bought loose are in some cases more expensive than the equivalent bought prepacked. Allowing consumers to select the choicest items means more waste, and in the same way that we can shop quicker by shovelling packs into our trolleys without overly examining the contents, so shelves can be stacked faster when the workers only need to slot in trays.

We have a choice, but a dilemma presents itself. Cheaper goods, but deal with the packaging, or buy the exact same item for more money but not have the packaging? The former has the caveat of producing more plastic waste, the latter of producing more food waste.

Perhaps developing biodegradable packaging is the preferred solution. Indeed, science has already taken up the baton and the search is on for truly biodegradable, sustainable packaging. But as there is no absolutely right path to tread, it is worth remembering that composting waste food in its biodegradable packaging will cause the production of biomass and the seemingly ever present CO_2.

Before we leave this topic, it is worth pointing out that not all households have the luxury of being able to choose to 'shop green'. We can't create a situation that makes fresh fruit and veg a luxury product or one that has to be passed over because of affordability. Maybe a concerted effort is needed. We need not be so fussy looking for the perfect specimen and supermarkets should stop trying to sell us more than we need. I'm not sure how plausible either of these solutions are.

Entice

Packaging can be produced in a variety of colours with all manner of labelling attached. Much has been made of the nuisance black trays so beloved of the supermarkets, which is nigh on impossible to recycle (not because the plastic can't be recycled, but because the recycling equipment can't readily identify the plastic as its black colour absorbs the scanning beams). So why use such an awkward colour?

It's all to do with psychology, a trick that brands use to convey certain images for their products. Black conveys luxury and sophistication; add gold lettering to it and we're hooked. Two identical items could be presented to us, but putting one into black packaging, and charging a little more convinces the brain that this is an expensive luxury item and engenders desire, so we buy. Have a look next time you are shopping: cheap cuts of minced meat for instance are quite happily sold to us in clear trays. Something like a premium minced Aberdeen Angus will come on a black tray (and we already mentioned the use of added gases to enhance the colour of the mince).

Of course, it's not just the colour of the packaging that is enticing us. Packaging is a nice handy way of carrying informative labelling, whether barcodes, expiry dates or product information. Fancy scripts or appealing pricing printed on a bright yellow background pulls us in. Being told this through the packaging enhances sales far more than leaving the product nude and sticking the information on the shelves. The enticing nature of packaging is a whole industry in itself, with many artistic and creative types hunched over design programmes seeking ways to make their product seem too alluring to just walk past.

This brings us into a whole other world of packaging that goes far beyond the supermarket shelves and into industries as disparate as top-end perfumers and grab-and-go blister packs. The design of the packaging is paramount to distinguishing the corner of the market best-suited for a product. We will happily pay more for a product dressed up in a fancy package than one displayed in a blister pack with a backing card. Strip away all the packaging and sometimes it is difficult to distinguish between products. Although we tend to discard the packaging, we do buy into the added appeal it gives a product.

A step too far?

Wherever it is applied and whatever the industry, the really annoying packaging is that which seems to serve no purpose. There is quite a bit of it about unfortunately, and again we need to head back into the supermarket for prime examples. Overwrapped turnips, cabbages and that old favourite, coconuts.

This leads us onto another packaging problem, suppliers prewashing vegetables such as potatoes, parsnips and carrots before bagging them. These vegetables, especially potatoes, used to be sold unwashed and in thick brown paper sacks. An odd stone would show up and a fair bit of clay. As packaging moved increasingly to plastic, these vegetables became presented washed. After all, they looked far better washed once they were on display in their clear bags, clearly potatoes and not muddy lumps. The problem is that many vegetables start to soften and spoil once they are washed. This of course reduces their shelf life. While it may have been good to eradicate stones from our purchased potatoes, the habit of washing is really not needed. Modern machinery can jig and shake surplus clay from our vegetables without harming them. We are perfectly capable of washing them ourselves, particularly if it means they won't go over so quickly. Most of us will wash veg anyhow as a matter of course before cooking.

Trimming vegetables is another 'favour' we get these days, but it too reduces the life of our veg. Cutting all the stalks off carrots and parsnips means they can sit snug and clean in their bags, but they will go off quicker. Likewise, top-and-tailing a turnip is doing no favour to the turnip but saves us from the rustic charms of its tapering root and leaves, and exposes it to the ignominy of a totally unnecessary plastic jacket.

All this washing, trimming and slicing means that we as consumers are increasingly being infantilised. Multi-packs may be one method of us selling goods without too much thought process beforehand, but this has now escalated to quite ridiculous levels. I'm talking here of sachets and increasingly smaller individually packed goods. We are given the message of convenience, but is it in fact all just getting out of hand? Do we really need to have our coffee spooned into individual sachets or our porridge pre-weighed in packets or indeed a handful of nuts in its own little plastic tub?

Before we leave this topic, there is one virtually unacknowledged portion packaging that has been around for a while and that is the teabag. Hardly noticed now, this packaging of tea for just a single mug leaves behind perfectly compostable, degradable tea leaves encased in a, for the most part, non-degradable mixed-material bag. Manufacturers are looking at alternatives, attempting to eliminate the plastic in these bags. Some have succeeded, but not all. Might it not be easier to use one of the many versions of balled tea infusers on the market rather than sling out mountains of teabags every year?

What about the rest?

Standing in an average supermarket it is not hard to spot the bottled water section, which in some supermarkets take up an entire aisle. I would argue that this is an unnecessary use of packaging and a needless waste of energy expended in the transportation of this water. It is commoditising what is a natural resource.

We are fortunate to live in a society where we can readily access clean and potable water at the turn of a tap. Where mineral or even spring water is desired then why not have that option? My argument is with the 'table' water which really is no more than standard water coming out of a tap at a factory and no different to the water coming out of your own tap other than that it may have been filtered before bottling. A great deal of supermarket water is just normal water with no special attributes.

Of course packaging is used more extensively than that found in a supermarket. Almost everything is packaged in one way or another, whether a new kitchen or a potato peeler, a wheelbarrow or packet of seeds. Most packaging nowadays uses a variety of packaging media. A pack of seeds may come in a paper envelope, but chanced are the seeds themselves will be in a plastic-film-lined foiled sachet. The paper is doing little more than carrying information, but the sachet is keeping the seeds fresh. Unfortunately it gives scant opportunity for recycling. The new kitchen will also tend to use a number of packaging materials and will also include cardboard, plastic and foams. As we know, only cardboard has much recycling use.

Years ago items were packed using natural products. Small items were packed in bran, then a waste product, now beloved of health and fitness gurus. Large items were placed in straw and the tea chest containing newspaper-wrapped china was ubiquitous when moving house. While there is no doubt that going back to the way we used to do things is not entirely possible or even desirable, maybe we can be a little more flexible in our approach. Maybe there is a use for straw in certain applications and perhaps even bran too. But using yard upon yard of Kraft paper instead of bubble wrap is doing no one any favours.

Paper is not infinitely recyclable. As we saw earlier with plastic, where multiple recycling loops entail an increasing breakdown of molecular chains, so it is with paper. The more paper is recycled the shorter its fibres become, until we are left with an unworkable mush. Paper comes mostly from trees, and, as we will see later, it doesn't tiptoe lightly past environmental issues on manufacture. While it doesn't hammer the environment like aluminium, for instance, it does come with its own problems. A multi-targeted approach is needed. When it comes to packaging, we need to consider all the facts about each material and its potential alternatives.

Recycling – the big question

We have already partly considered recycling in the last chapter, but it's worth expanding upon now we have seen a bit more about how plastic is used in packaging.

To recap what we have discussed so far, much plastic as we know is eminently recyclable, but not all recycling equipment can decode its characteristics so it's discarded. Likewise there is absolutely nothing intrinsically wrong with black plastic. Most of it is PET, a valuable material; it's just that the technology needs to catch up so it can be identified correctly it when it enters the waste stream. Some film can theoretically be recycled, some is problematic as it comprises many layers that can't be easily separated or gets tangled in machinery and does not shred like bulkier plastic.

What is and isn't recycled in the UK is another issue and one in part that relates to economics. Councils and recycling centres may well pick over the choicest items and direct their recovery focus there. A check of Waste and Resources Action Programme (WRAP) prices[26] may show where the greatest value lies. In 2020 unprocessed or virgin polyethylene costs around £1200 per tonne (HDPE slightly more, LDPE slightly less), while the price of uncoloured HDPE once recycled comes in at £490 per tonne. The price if tinted drops to £385 per tonne and indeed has through 2022 dipped substantially lower. HDPE is used to make milk bottles. Take off the lid before chucking it in the recycling and you add to its value, although it is worth reiterating, the advice now is to leave lids on these bottles.

These figures remained characteristic for some time but at the time of writing they have become somewhat erratic due to a number of external driving factors. Most notable are the price changes occurring for PET. Buying the virgin material will set you back around £1325 per tonne. Recycled, the clear PET as used in these endless water bottles used to cost in the region of £222.50 per tonne; as soon as a colour is added (even through just not removing the lid), that price plummets further to £50 per tonne. However in the summer of 2022 clear recycled PET jumped in price to around £1,478 per tonne, higher than the virgin price. This unusual pricing reflects the tax imposed on packaging that contains less than 30% recycled material.[27] This tax, coming into force in April 2022, dictates that where the 30% recycled content is not met a fine of £200 per tonne is imposed. This drives the price since supply is not meeting demand and the tax makes it worthwhile paying inflated prices.

Councils' restrictions on 'hard plastic' can add yet more confusion. This can include anything from flower pots to drain pipes to toys to any number of oddments that turn up in the average household. The difficulty in identifying them is one problem foreseen by the recyclers, as is the possibility that they comprise blends, fillers and additives that just make it too difficult to lump in on one recycling stream. Issues may be practical and polymer-related. PVC, for instance, can be problematic. It is a very common plastic and has a long history in packaging, but it does have specific processing demands. If not processed at the right temperatures and in the right conditions, it can break down, releasing hydrochloric acid.

A whole host of other issues come to play where plastic film is metallised, a process that is beloved of manufacturers of crisps and chocolate, as well as gift wrapping and hydrogen balloons. Likewise problems come into play with the increasingly rolled out 'solution' to plastic ready-meal trays of replacing them with paperboard. Unfortunately this paperboard has to be coated with a layer of plastic film to stop the tray contents seeping through. If plastic film is difficult to recycle, imagine the difficulties faced if it first has to be scraped off a paperboard base. It's not just found in microwavable trays, this coated paperboard is common in sandwich packaging and salad trays also.

Just because a product has a recycling stamp on it does not mean that anyone is willing, or even able, to recycle it. We can't bury anything problematic and forget about it, nor can we ship it abroad, where experience shows us that it is not always dealt with as we hoped. Scientists and product developers are working on potential solutions. One such relates to a device that can be fed scrap plastic – whether bio-based or hydrocarbon – which on pyrolysis yields fuel. The pyrolysis process itself, however, requires great deal of energy, but if this can be overcome then such a device would make it profitable for populations globally to recover and recycle plastic waste. This is probably some way off, but possibilities are out there, with great reward for the person that can effectively crack the problem.

In the meantime, raw material suppliers are also looking for solutions, not only by developing new properly biodegradable plastics but also by modifying the formulation of existing plastics. One area of investigation relates to the bonds holding plastic together. If you recall our conga line from an earlier chapter, polyethylene for example consists of bundles of hydrogen and carbon, all joining up to form a long chain. Using our analogy each bundle frees their 'hands' and holds their neighbour. Now imagine that instead of a tight grip, this is more like a limp handshake that could easily be broken. If these bonds can be broken post-use, then instead of being left with unwanted polyethylene, we can be left with just ethylene.

Packaging the packaging

So to sum up the supermarket packaging issue. Some of it is needed. Long term it really does help keep our food safe, longer-lasting and hygienic. Reducing food waste as much as possible has got to be a good thing from all sorts of angles ranging from environmental to moral. But there does need to be an effective way of dealing with that packaging post-use.

That leaves the mountain of plastic that swathes countless items from prewashed salad to sliced courgettes to destalked grapes. We know it's difficult to deal with from an environmental perspective. It for the most part can't be recycled in conventional systems because, as of yet, a lot of recyclers equipment can't deal with it. Being lightweight and thin means it has a tendency to get caught in the reprocessing systems.

So do we point the finger at the plastic manufacturers since they produce it? Well no, they aren't going to turn down work. Anyhow, if one refused to produce it there would be plenty others ready to fill the void. Is it the fault of supermarkets for giving us a solution to a problem that never really existed, at least not to the scale that is currently marketed? Again economics comes into play. There are shareholders to satisfy and profits to be made. If there are customers willing to buy then they will keep churning out product and become ever more inventive. If they thought there was a market for pre-squeezed toothpaste on disposable toothbrushes, then that's just what would be filling the shelves. That leads us to the customer. If we as consumers stop buying certain excessively packaged goods, they will stop supplying. The power is in our hands.

8

CONSTRUCTION AND BUILDING

The construction and building industry consumes a hefty 20% of polymers produced annually. These can be found in anything from damp-proofing to roof tiles, insulation to windows and doors, and all manner of conduits, profiles, seals, guttering and pipes, even down to the paint on the walls.

Much of the plastics used in building and construction are easy to spot, but a great deal more are not so apparent. It will be interesting to take a tour of the average house to investigate just where and indeed why plastics are used as they are. But before entering the home, let's first take one specific and important area of use within the construction industry, and that is pipes.

Pipes

This is a particularly important area, as it consumes in the region of 35% of the plastics used in this sector.[28] Pipes are very versatile and can be used to carry all manner of liquids or gases, basically anything that can flow, including water, gas, sewage or oil. They can be hidden underground, span miles, crisscross cities or frame houses as guttering and downpipes.

Back in the realms of history, pipes carrying water, particularly those in London from the fifteenth century onwards, were made from hollowed elm tree trunks. Elm was chosen as it didn't decay when kept wet, making it particularly good in transporting water.[29] It worked well and remained in use until the eighteenth century. Of course there was a downside: a lot of elm trees were needed keep up with the ever-increasing demand for piped water, especially as towns and cities expanded.

Even older and spanning a greater historical time period is clay piping. Their usage stretches from prehistory right up until relatively modern times. Terracotta can still be found in pipe manufacture, but these pipes are very heavy and problematic when it comes to transport. They are also prone to attracting damaging tree roots when buried underground and are susceptible to leaks, cracks and breaks.

Metal cast-iron pipes were used from the seventeenth century, and in some cases are only now being replaced. These iron pipes can be found deep underground and generally carry fresh or waste water. Lead pipes are far older and were used from Roman times to transport water, especially domestically. In fact, so synonymous are lead pipes with plumbing that the very word 'plumbing' is derived from the Latin word for lead.

The UK government banned the use of lead in many areas of application, and by 1970 it was no longer used for water pipes. Up to that period these pipes were found in a domestic setting in the form of guttering and downpipes or carrying water from larger iron street-dwelling pipes into the home and direct to the tap. Needless to say, lead cannot be used in this capacity now. Indeed, both lead and iron give off particles of rust and, in the case of lead, other toxic elements can also be taken up by fluid passing through it. Copper is also used in the home and replaced lead for water piping; it can also be found in central heating piping. In this application, plastic replacements have meant far easier installation, with flexible fittings negating the need for bends and joints and solders (also formulated from lead up until relatively recently).

Plastics usage in piping has been stepped up with programmes nationwide looking to replace rusted old Victorian metal pipes with lightweight, durable and easy-to-handle plastic pipes. They can be seen looking like giant insect hotels in their wooden-framed cages stacked up on the side of a pavement, each colour-coded and awaiting deployment. Medium-density polyethylene (MDPE) is the material of choice here and used extensively for underground piping. This plastic is of the same basic type as many a plastic bags or milk bottles. The medium-density bit just means that their molecular chains are packed a little closer than found in low-density (LDPE) bags and less close than high Density (HDPE) milk bottles, for instance.

MDPE is useful in many areas (more than just piping) and is colour-coded so the pipes' contents are easy to determine. So next time you peer into a hole in the road and see these pipes you will be able to say with authority that blue MDPE pipes are used for potable water, black pipes are used for waste water and yellow pipes for gas. MDPE pipes can also be coloured black with two orange parallel lines running down its length. If you have anything to do with interred pipes it is worth keeping a look out for this striping. These are the pipes you don't want to accidentally puncture, as they transport sewage or industrial waste.

Fortunately, one major advantage of plastics here is their ease of colouring. The colour can be incorporated permanently, so there is no chance tell-tale orange stripes can rub off and contents misinterpreted.

Construction materials

Plastics have a myriad of uses in construction apart from the aforementioned piping. If you have ever kept dry walking from a supermarket car park via a covered walkway, or spotted children playing under an outdoor canopy, stored a pushchair in a covered parking area, grabbed an al-fresco cigarette break, stashed a shopping trolley or waited for a bus under a glazed shelter, chances are these were constructed from polycarbonate (PC). PC is a great glazing material. It has terrific transparency, is incredibly strong and has very good impact resistance (which is why it is also used to make police riot shields or those more robust safety glasses sported with increased regularity on television programmes where debris or certain chemicals may be a hazard.).

PC has many positive attributes, but is not great in the presence of hydrocarbons, such as petrol, which may be why some roadside bus shelters appear a little crazed after a while.

Moving from the kerbside and into the building itself, plastics surround us, sometimes subtly and other times blatantly. Plastics can be found within the walls in the form of insulation material and in wall linings. Insulation comes in many forms, from thermal to acoustic, fire to electrical (this is internal insulation as opposed to the more visual external panelling). Polymers are very useful in these applications as they either have inherent qualities that make them applicable, or they can be imbued with additives that enable them to do the job required.

There is a very simple method to show how thermally conductive a material is: just touch it. Of course this is best done if the materials in question are at room temperature. What do you feel? Is the item warm or cold? If you are touching a metal, it most likely feels cold. Glass and ceramic give this illusion of coldness too. Now touch an item made of plastic – is it warm to the touch? It is all to do with the second law of thermodynamics and finding equilibrium, but we'll keep this in simple terms.

Touching a metal at room temperature – that is at around 21°C – with your hand, which is nearly double in temperature, means that there is a flow between you and the metal. The metal is drawing off your heat in an attempt to equilibrate or balance the temperature. This thermal flow feels cold to the touch, as your heat is being drawn away from you.

The better the conductivity the faster is this flow. Metal is a good thermal conductor, so you immediately feel metal to be cold to the touch. Plastics feel warm. They are good thermal *insulators*: the subtle warmth you feel is a reflection of the heat in your own hands transmitted back to you. It is why for instance we had steering-wheel covers years ago; my parents had a natty knitted version. Steering wheels were metallic so were freezing in winter and too hot to handle in sunny summer months. Now steering wheels invariably have a polymer outer layer and stay pretty much at an accessible temperature, so we have dispensed with covers or driving gloves.

Foamed materials also sit within the repertoire of insulating plastics for building use. They can have very interesting properties, especially when considering that they are no more than the base material and air. Yet air can be an even better insulator than just plastic alone (which is why your mother told you to wrap up in layers to trap the air and keep warm). Combining air and plastic makes for a very effective synergistic insulating material. This is why expanded polystyrene (EPS) has been used so much in this arena, where it can be seen anywhere from lofts to cool boxes. Polystyrene lends itself very well to foaming, is cheap and has good thermal properties. EPS fares well against traditional insulating materials such as fibreglass and is a little better than natural materials such as sheep wool.

Keeping an ever-present eye on environmental issues, this is another instance where certain choices can be made in terms of renewable natural sources. The use of animal products and synthetic materials, including the likes of glass, can be very effective, but do have disadvantages as well. So if the environment is a concern and the sole reason for excluding plastic as a possible solution is just the fact that it is plastic, then we must realise that all options come at a cost to the environment. Even wool. This will be explored a little later, for now the attributes of plastics in insulation application can be explored a little further.

Insulation as used in construction can come in many forms from beads to board, foamed blocks to foil-backed matting and many in between. Foams are useful as they can utilise the synergistic qualities found in both air and plastic (as we already mentioned). These can be enhanced further depending on the form of foam employed and this involves the bubble structure of the foam itself. Foams contain structures that can be described as open or closed cell.

Open-celled foam is the squidgy type which can quickly bounce back when depressed. Here the individual cells of the foam are linked together forming a tunnelling effect through the foam. Air can pass through, but not so easily that heat is lost. A typical example of an open-celled plastic would be polyurethane. This is polyurethane in the form of soft sponge that can be found anywhere from its position beside the kitchen sink waiting for a dollop of washing-up liquid, on the side of a bath or as settee cushion inners. This isn't the only example of open-celled insulation materials. Natural materials such as animal or mineral wools can form the same type of structure. Just think of a loofah – this too has an open-celled structure.

Closed-cell foams are just as inferred. Here the expanded cells of a material are gas-filled but each are self-contained, trapping the gas, and so don't connect with each other. In this case think of a bubbly chocolate bar. Individual bubbles can easily be seen through the cross-section of the bar with each bubble distinct and separate from its neighbour.

In plastics terms EPS is a well-known example of a closed-cell foam. Polyurethane (PU) somewhat confusingly can also be made to be closed cell. It all depends on the manufacturing method used. Closed-cell PU can be found in the rigid foams used in many insulation applications, including structural panels or the insulation squirted around a window frame when setting it into a building.

The size of the bubbles within the structure can be altered, and this also affects its properties. Some brands produce small, tight-packed bubbles, while others are larger and create less dense structures. Smaller rigid bubbles can provide better protection in some applications, when sealing window frames for instance, than its larger-celled counterpart, used where bulk is required.

Closed-cell insulation can also include spray foams, another example of the ever-versatile foamed polyurethane deployment. Here the foam is sprayed before it cures (PU is an example of a thermoset plastic – i.e. one that hardens via a chemical reaction), where it then sets in situ.

Moving on from insulation, another interesting and maybe unre-marked use of construction plastics comes in the form of archi-tectural fabrics. Architectural fabrics can be seen in many modern structures, perhaps most famously in London's Millennium Dome/ O2.

These are the structural elements that allow the building to swoop or soar, constrained only by the architect's imagination. These aren't necessarily load-bearing components designed to support multiple storeys, rather they are the finishing outer structures that impart the aesthetic artistic flourish, making the building memorable.

While all manner of cabled and internal structures hold the fabric, the large surface area fabric itself can be made up of a number of materials from cotton to PVC-coated polyester (giving a double whammy of plastics). This latter choice is the more common option, as it offers a good life span and is reasonably inexpensive. Of course other plastic and natural materials (as well as all manner of combinations) abound.

In fact, coated fabrics can impart all sorts of additional advantages. Coatings of PTFE can give similar non-stick properties to those conveyed by coatings on the inside of a frying pan. Used on buildings, whether layered onto these fabrics or as a thin coating on glass, it will result in a self-cleaning structure. Dirt and indeed all manner of liquid, including that excreted from a passing seagull, will struggle to find purchase, thanks to its low-friction surface.

Another more recent and well-publicised application for these coated fabrics can be seen in the construction of football stadia for the 2022 Football World Cup in Qatar. Despite the tournament taking place in November and December, football fans and players were thankful for the PVC coated fabrics utilised in the roofing structure. These fabrics with, their additional layer of aluminium pigment, were designed to reflect the sun's rays away from the stadium so reducing heat build-up within the structure by limiting heat radiation.

Moving onto another oft-overlooked application, have you ever noticed large sweeping timber trusses spanning a building, appearing as modern timber framing or as majestically sweeping beams curved to mimic the undulations of the structure it supports? These to all appearances seem to be nothing more than solid timber beams. Yet beams of such size and strength must be difficult and expensive to source. Or maybe all is not as it seems.

A modern addition to an architect's portfolio of building materials is glulam, a contraction of the term, 'glue-laminated timber'. They comprise multiple layers of thinly sliced solid wood bonded together with a strong structural polymer adhesive such as acrylics or cyanoacrylate (a glue that is not a million miles from a superglue). Basically they are like a structural form of plywood, but more closely resembling a solid timber beam than a sheet. They are very strong, are moisture-resistant and can be very aesthetically appealing, which is a strong selling point.

These glulams can be found straddling roof spans, providing framing to a house or spanning decks. They are now widely used as in appearance they seem to be on-trend solid beams of wood, yet by clever laminations they can be made longer, far stronger and incorporate more complex-shaping than traditional timbers. In environmental terms, glulams can be appealing, as they allow for far less wood to be used in construction. The traditional construction-friendly species of larch, oak and spruce are employed, but these beams are built up of many layers rather than solid timbers. This has several benefits. It means for one that beams can be made to the dimensions required, so there is less wastage by hewing away timber, but it also means that timber otherwise rejected because of knots or other defects can still find a use as part of the laminate.

This can make for a happy marriage, a natural product made more effective and efficient through the use of a polymer. Yet it is done so with a subtlety that hides the very presence of the synthetic material. After all, if everyone wanted to go back to solid timber houses we would soon return to the Tudor problem of severe wood shortages.

Plastics crop up in several other locations within the construction industry. Anyone viewing a building site will notice swathes of it. Some is present as packaging in the form of large-scale overwrap, smaller pouch packets or blister packs. Some appears as large rolls of damp-proof lining, soil-retaining mesh, seals, underfloor heating paraphernalia or protective sheets on glazing products and, at the other end of the house, synthetic roof tiles.

All manner of fiddly little gizmos and clever devices indispensable to the constructor abound. These can encompass such disparate items as tile spacers, clips for pipes or wires, ties, wedges and a whole host more, including stopcocks, brackets, levers and conduits. Even in large-scale construction sites odd little plastic devices show up, including those colourful little caps pushed onto the end of steel cables sticking out of reinforced concrete. These mushroom caps handily stop constructors inadvertently poking themselves in the eye with the exposed cable end. And what about rubble buckets, chutes, scaffolding mesh, barrows and barrels? And of course we can't forget those present-day essential accoutrements to any building project, the hard hat and the hi-vis jacket. Like so much else mentioned here, these have a good chance of being manufactured from that reliable standby, PP.

A tour of an average house

An indicator of our indebtedness to polymers comes on considering their use in a domestic setting. Strictly speaking these aren't all construction and building applications; nevertheless, this exercise is useful in demonstrating where plastics can be found in a typical home environment. Uses are so manifold and varied only an overview is needed.

One method of considering polymer use in the average house drops us metaphorically at the front door. From this starting point we can tour the house and begin to consider our homes with fresh eyes. It soon becomes apparent that it may be easier to consider where polymers aren't used. Let's pick out a few interesting highlights. This can begin right at the door from the paint on the walls, and the varnish, laminate, lino or even carpet on the floor. Even a wool or natural fibre carpet can have polymer based backing or underlay.

More obvious are the seals, beading and fillers surrounding doors and windows that may in turn be PVC based. Our hands can stray to wall mounted sockets and light switches (ABS or maybe melamine). Behind the scenes are a myriad of wires swathed most likely in PVC (but not exclusively so).

Getting comfortable on any padded chair, settee or even bed means we encounter a cushiony combination of materials. Polyurethane (PU) foam will most likely be found whether formulated to have open or closed cells (dictating the firmness of the foam padding). PU might form the main foam constituent but can be layered with tougher foamed ethylene vinyl acetate (EVA). Foamed EVA in turn can be found in a multitude of applications from gardening kneelers to acoustic dampers to those foamed inserts that hold say, a new drill in its box.

Room by room polymers each have their uses. Sometimes familiar names crop up across several applications and sometimes specialised materials are deployed in specific location. The business end of a lamp or ceiling light for instance will use high temperature resistant materials such as the amino resins or polyphenylene sulphide (PPS). This is that dark blackish brown material that the lightbulb screws into. Otherwise acetal (POM) is generally used where strength, toughness and resistance to wear is needed. Nylon (PA) offers good strength, toughness, temperature resistance and thermoplastic polyester (PET) – as opposed to the resinous thermoset– is strong and stiff, gives a high degree of dimensional stability and a broad service temperature.

Before we finish this tour lets pause briefly in the kitchen. Here polymers can be seen in an array of applications. What about the knobs, handles and door facings on the cooker, saucepan lids, cupboards, dishwasher, washing machine? These all most likely will have encountered plastics somewhere in their make-up, some more than others. The insides of the dishwasher and fridge in particular will throw up a further array of functionally dependent polymers.

And finally the bathroom where silicone sealants are used in abundance keeping leaks in abeyance. From the toilet seat (which has come a long way from the hygienic black phenolic toilet seat once the staple of many homes), to the electric shower and cubical, to the bath, all are dependent on polymers in one form or another.

Now look back into the house and image it stripped of all plastics. It would look markedly different with much lost to us.

9

ELECTRICAL/ELECTRONIC

Plastics have played an instrumental role in the rise of the electrical and electronic sector. In fact, their concurrent rise to prominence is a notable reflection of the dependency this sector has on plastics. While the science of electricity has been explored since antiquity, when pieces of amber (a natural plastic) were seen to possess unconventional properties, it was during the mid-nineteenth century that greater investigation and forward strides took place. While both were still at an embryonic stage, advances in telecommunications (as seen in an earlier chapter) was made possible by gutta percha, This rubber-like material enabled electronic pulses to travel through cabling laid across the globe, particularly underwater. It was in many ways a nascent world wide web and certainly was a hugely contributing factor in enabling electrical communication.

Successive generations of polymers have made great strides possible within the sector, and now we enjoy all the advantages of a vast array of electrical and electronic goods. Indeed, it is difficult to consider any piece of electrical or electronic equipment that does not contain integral quantities of polymer. Strip back a mobile phone and a complex array of materials are uncovered, polymers included. From cable sheathing to appliance housings to epoxy circuit boards, polymers are essential to the vast collection of equipment we both rely upon or simply desire for what they can allow us to achieve.

In general consumer electronics can be broken down into colour categories. Some will have a familiar ring, while others may not be in common parlance. Probably most will have heard of white goods. These, as implied, refer to those appliances that traditionally used to be encased in white enamel, and include fridges, freezers, air-conditioning units, washing machines and dishwashers. While these goods are still likely to have a metallic or enamelled shell, so many elements contained within are polymer-based, from knobs and handles, to liners and drawers, and that's before the little bushes, gears and cogs are accounted for.

Small white goods refer to various essential items found dotted about the kitchen, items that again traditionally may have been white. These include toasters and kettles, blenders, choppers, food mixers, coffee makers and microwaves. Anything really that will sit on a counter top. Items like vacuum cleaners or hairdryers may appear here or be put under the category of brown goods.

Brown goods are those items that traditionally would have been housed in wooden boxes. This could be anything from the television set, originally shrouded in a solid wooden cabinet, to the record player, which in our house used to reside in a stylish wooden cabinet with a lift up top and spindly black wooden legs. Now a great number of goods can sit under the label of brown goods, from the television and accompanying remotes to audio equipment, CDs and computing systems.

Once more, there is a little overlap with another colour-coded branch of electrical appliances which this time comes under the heading of black goods. These are probably replaced or updated with greater frequency and comprise communication or recreational goods. As can be guessed, these would originally be found in black casings. Many are personal electronics such as mobile telephones, personal stereos, headphones or even laptops and tablets.

This leaves us with electrical tools, which can be domestic or the more robust professional variety, and office goods. With the latter there is a little overlap with that found domestically, although nowadays the differences are shrinking, especially with the now commonplace personal printers, computers and other office paraphernalia. Again much has been covered in previous chapters, but it is not difficult to find plastics in these applications. Looking at a keypad, whether on the computer or telephone, it is apparent these are plastic, sometimes with softer rubbery silicone keys that yield in a satisfying manner when depressed, or harder and offering greater resilience as in the tappy ABS keys on a laptop (these, by the way, are made of the same materials as the Lego brick). Reliance on polymers in office goods is not hard to uncover, from paper support levers to ink printer cartridges, from the mouse to the multitude of USB devices.

All this is still only considering the smaller end of the scale. There are vast electrical infrastructures dotted about the country, from factories, to power stations, to the large data centres needed in the modern world.

Data storage is now likely to be offered to businesses and consumers as a virtual entity and not physically found as a server within a given premises. But storage capability does need to be held somewhere tangible. Now there are data storage centres which contain the physical infrastructure necessary to hold data and allow cloud access. To function efficiently these centres need not just the servers, but a number of peripheral components including for instance routers, switches, firewalls, storage systems, servers and application delivery controllers.* Needless to say, such systems can only operate as they do with plastics in their construction.

Reliance on plastics

The reasons for so much plastic usage in this application are manifold, encompassing cost, aesthetics, durability and of course, insulation capability. In terms of cost, it can be far more economical to manufacture a drill, say, from a plastic than metal. This goes beyond the base cost of the material and reflects also the ease with which plastics can be formed into complex shapes. Electrical goods such as the aforementioned drill (but true for so many other goods too) first emerged onto the market encased in metal. This casing tended to comprise several disparate pieces of moulded metal that needed fixing together. Assembly could be achieved via welding, riveting, gluing or welding. While similar assembly methods are available with plastic parts, there is on the whole less need for fixing. Plastics are capable of being moulded in such a way that several parts can be incorporated and moulded as one, so vastly reducing the number of separate pieces involved in manufacture. Plastics can also be coloured, and any such colouring is embedded in the material and won't scuff or chip away.

An added bonus of using plastic is its durability, and this of course is a reflection of the actual plastic used. Mention has already been made of the stalwarts in this and many other application areas in previous chapters. It is then no surprise to realise that ABS, polypropylene (PP), epoxy and nylon are commonly found here.

As before, ABS comes into play where a robust yet glossy appearance is required. This includes areas such as keyboards and the housing of gaming paraphernalia, telephones, kitchen gadgets and tool housings.

Polyolefins, which include PE and PP, are very apparent in this sector and show up throughout the industry as wire and cabling sheaths and insulation, amongst many other uses. PVC is also very prominent in this area and has been for a number of years.

PP can also be found in applications where its inherent high electrical resistivity is needed. This together with its low coefficient of thermal expansion (i.e. it doesn't excessively expand when heated or, in other words, it remains stable) gives PP an advantage where resistance and durability against heat and electricity are needed.

Aesthetically, PP also comes up trumps in such applications due to its ability to exit a mould replete with a glossy sheen. This sheen can in part be attributed to the inherent characteristics of the material, but the role played by the mould itself needs acknowledgement. A high shine on the interior of any mould deployed will mirror onto the surface of the resultant moulding.

Of course, modern electronic technology is heavily dependent on plastics, not least the ever-present mobile phone. Here the casing will probably be ABS with a polycarbonate (PC) screen – some models will use PC for the entire casing. One advantage undoubtedly is the aesthetic value of using plastics. These cases can come in a variety of colours. Other plastics can also be found in phones including nylon and acrylic. Mobiles are compact powerhouses with a remarkable range of materials stuffed into such a small space. Many are dependent on each other for the device to work. Plastics are integral to this, while also contributing to the production of a lightweight device able to reach the market at a lower cost.

Close proximity to electrical currents usually involves thermosetting materials (although others can be used in certain instances), as they tend to have much higher operating temperature ranges and won't melt if they overheat (if you remember thermosets are chemically fixed and won't readily lose their integrity when exposed to elevated temperatures). These include the amino resins such as melamine or urea formaldehyde, phenol formaldehyde or epoxy. These can be found used in areas such as light fittings, switches and fuse boxes. They can also be found on saucepan handles or cooker knobs, where high temperature resistance is also important.

Urea formaldehyde (UF) is worth singling out, as it is particularly advantageous in all things electrical, performing extremely well not just when exposed to heat but also when in contact with a range of voltages. UF's ability to withstand high voltages allows it to provide a high degree of electrical insulation.

Indeed, this material and others, such as thermoset polyester, are widely used in electrically insulating applications. Where once porcelain insulators were deployed when high voltage was encountered, it is now increasingly likely that insulation will come from a polymer material. Such is the capability of these polymers that they can now be found in anything from industrial power conduction to circuit-breaker components and housings, electric motor brush holders or even the housing for the bimetallic strips that automatically switch off a boiling kettle.

Many of the workhorse polymers integral to making our electronics work as they do go about their business efficiently and for the most part unnoticed. Their names may be unfamiliar and their capabilities unacknowledged, but without them our modern world would not function as it does. It may be worth name checking just a few to give an essence of their usefulness.

Polyphenylene sulphide and polysulphone are high-performance thermoplastic materials and as such have above average capabilities when it comes to heat and electrical resistance. They tend to be used where a great deal of heat can be generated such as at the business end of a hairdryer, steam iron, toaster or even microwave. These components consist of items such as the grilles that allow heat to escape on a hairdryer, the turntable key or grille of a microwave and the valves and switches of an iron or toaster.

PTFE, a.k.a. polytetrafluoroethylene, is well known as a tape or by its trade name Teflon where it coats cookware for its non-stick qualities. PTFE is also used in electrical applications and is a very good insulator. Other materials such as EVA we have come across in the previous chapter. It is used both in moulded and foamed form and can be found in the insulating strips on fridge or freezer doors, soft grips on electrical goods such as for instance an electric toothbrush, or hoses or tubes in domestic appliances.

A host of other plastics, from both ends of the price range, are used also in all manner of applications. Poke away at any gadget, whether large (like a dishwasher or fridge), or smaller (such as a toaster or coffee machine), a broad range of plastics can be found. Internally, nylon and acetal, externally, polystyrene or polyphenylene oxide, and here we are only looking to the kitchen for examples.

To sum up, plastics used in electrical applications tend to vary between those that are very much in evidence, such as the housings of appliances or gadgets, and the high-performance elements mostly sited internally where voltage or high temperatures are encountered.

Generation of electricity

Energy harvesting is an interesting area undergoing sustained investigation and currently studies are looking at the many and diverse ways it can be achieved. Rubber comes into its own in many of these studies which look to draw energy from a common action or motion. The basic principles are very simple and take their inspiration from the manner by which static electricity can be generated. So for instance if you have ever rubbed a balloon on your hair and found it stuck on a wall, or rubbed a glass rod or even a piece of amber with a cloth and found it could pick up small pieces of paper, then you will know about this form of energy generation.

From this starting point studies are taking the core concept and devising ways in which energy can be harvested when two dissimilar materials (one of which is normally a rubber) come into repeated and cyclic contact with one another. This can be as simple as having the two materials pressed together and pulled apart repeatedly. This is called triboelectric energy harvesting. The concept has been around for a while, but now studies are advancing to show how this can be made both practical and useful outside the lab.

Rubber, as mentioned, comes into its own here. It is a very useful material to explore and is yielding some very unusual harvesting possibilities. For instance, why not harvest energy during everyday activities? One way of achieving this would be to embed devices in our shoes that would harvest energy each time we took a step. This would work by using rubbers (like those in a balloon) that would sit within the sole of the shoe. When the shoe is flat on the ground the rubbers abut; flex your foot to take a step and the rubbers come apart. Repeat many times as you walk to the shops and this continuous movement is in effect kinetic energy. This kinetic energy has the potential to be translated into electricity. There is not enough to power your home of course, never mind the impracticalities of getting it into the grid, but there could be enough to, say, power wearable tech or charge a mobile.

Similarly other movement can be captured and likewise used to generate low-level electrical current. Little devices could be implanted in the pavement, say, and as foot traffic traverses energy could be used to help power lighting. Rotary action such as the movement of a car or even bicycle wheel can also be used to generate this form of static electricity. This would be in addition to the dynamic energy that can be harnessed from the axle rotation. This isn't too dissimilar to the little dynamo lights found on some bikes.

This isn't so farfetched or futuristic as it may sound. Agencies such as those overseeing the highway or airports, particularly in the USA are looking at just such energy harvesting and are looking to integrating ceramic/polymer composites into roadways, rail tracks and even runways to generate enough energy for lighting, signage and traffic control. London too is among many cities looking to explore this area of energy harvesting, with Transport for London (TfL) trialling its 'intelligent street' concept just off Oxford Street.[30] As part of this trial they are investigating power derived from footsteps. Much needs addressing, such as the source of these materials as well as their effectiveness and longevity (for instance what is the balance between energy harvested and energy expended in producing the devices), but the potential is there and it is certainly an interesting concept.

When talking of these developments, we are looking in the main at low level electrical generation. Here polymers are used in conjunction with other classes of materials. Of course polymers are also used in the mass generation of electricity also. In addition to larger conduits, pipes and covers, this usage tends to be particularly prevalent in all those little switches, knobs, relays, circuit boards and so on already mentioned and so this point does not need to be laboured over again. Indeed pretty much all methods of generating electricity utilise polymers somewhere in its infrastructure. This really does encompass anything and everything from hydroelectric plants to nuclear power stations.

Polymers, maybe somewhat counterintuitively, are used quite extensively in the generation of green energy also. Solar panels, mentioned previously, are probably lower down on the scale of plastics usage. Wind power however is heavily reliant on plastics where it is a key requirement.

Much investment is going into harnessing wind energy. These are several iterations away from the picturesque wooden windmills of the past. Now gigantic turbines stalk across the land and not stopping there, now stride onwards into the sea.

These turbines are enormous with blades that can whip through the air at upwards of 200mph at their tips (dependant of course on wind speed and the length of the blade although this really is the upper limit at which they can operate). While the turbines on the whole are constructed from an amalgamate of materials (according to the National Renewable Energy Laboratory by mass this includes steel (71–79%), reinforced plastic (11–16%), as well as iron copper and aluminium), the blades themselves are most likely to be reinforced polyester.

Although difficult to appreciate on gazing upward to the lofty heights they inhabit, these blades are cleverly designed to make the most of wind energy while ensuring they can withstand the relentless weather, gravity, cyclic loading, bending forces, air density and all those other difficulties encountered throughout its functional life. These blades can be truly colossal, with lengths typically in the region of 115-150ft (35–45m). A new supersized turbine due in 2024 will boast blades of 380ft (116m) in length. Just looking at the 'smaller' blades currently deployed, we are still looking at spans equating to the length of a Boeing 737, which really is enormous for something designed to whip through the air atop spindly towers of steel.

To build a structure of this size vast quantities of raw material are required, not least in the huge amount of plastic needed to form the blades. Originally utilising steel, aluminium and even wood, it was soon apparent that great weight savings and greater ease in fabricating could be had if glass-reinforced plastic (GRP) was used instead. Now it is most likely to find composite materials making up these blades, where the composite can be the highly effective GRP, carbon fibre-reinforced plastic (CRP), which can make even stiffer lighter blades, or other non-glass composites, where the fibre component can be another polymer (such as aramids – think bulletproof jacket material). In each case the matrix, or binding material, will be a polymer, and this polymer will almost certainly be a tricky-to-recycle thermoset, usually either epoxy or polyester.

With the blades,[31] the important factor in their manufacture is to ensure that appropriate strength and stiffness properties are imparted in all the right places. In the early days, this would involve layering up sticky polyester or equivalent resin with long lengths of glass fibre in open moulds. Each layer of fibre would need to be completely coated in the matrix material and air bubbles eliminated. This was achieved by slapping the matrix on with a paintbrush or roller. When dry the shell halves would be stuck together with a suitable bonding adhesive.

This method was not without its problems, not least of which were the volatiles emanating from the matrix. As the blades grew longer this became more problematic. The labour intensity of such a method could not be overlooked, nor the difficulty in maintaining a given quality standard. The process became automated. Now there are a number of ways in which a blade can be manufactured, again to an extent depending on the manufacturer involved and the process favoured. For instance one method involves something called 'Pre-peg'. This refers to a composite material having fibres (whether glass or carbon, or something else entirely) pre-impregnated with the chosen matrix. These are then bonded together and formed most likely into a sheet form. This is then easier to handle and easier to reform into whatever ultimate shape is needed. Once shaped to the required aeroshell contours, it then undergoes a curing process which sets the form permanently.

Another process and one which is most commonly used for blades of these dimensions is a resin infusion process, one form of which is known as Resin Transfer Moulding (RTM). Here the resin, again polyester or epoxy, is injected under pressure into a closed mould. This mould would already contain long fibres of glass or carbon or maybe an aramid. Clever alignment of these fibres can ensure great strength and stiffness in very particular spots (directional strength) along the blade, maybe where bending forces are greatest, or maybe where storm force winds are likely to hit. It can all be quite closely engineered. For turbine blades the majority of the fibres however align unidirectionally, that is to say, all along one plane or direction. Curing comes after the injection process, and again this acts to permanently set the shape into whatever form the mould takes.

While the turbine is in service, this plastic is useful and performing an important function; however, consideration must turn to post use. After all, withstanding storms, frosts, intense heat and all those other elements a blade must encounter, will take its toll. They do have a fixed life span and many of the blades put into operation during the 1990s are reaching the end of their proscribed life. So what happens to the blades post-use? For the metallic elements recycling is, by and large, straightforward, but as the blades incorporate those tricky to recycle thermosets, the solution may not be quite as clear. In many cases old turbine blades lie stacked in blade graveyards (rather like the infamous tyre dumps) or landfilled until someone can find a use for them. However some lateral thinking is giving turbine components a second life.

One solution is to utilise the streamlined and massive form of the blade to repurpose them into functional new forms. For instance there is a city in Holland which has come up with the novel idea of using old blades to make slides, tunnels and ramps for a children's playground. The Dutch are also thinking laterally with plans afoot to design a bridge (earmarked for Denmark) using old blades.[32] This second-life approach for decommissioned blades is giving rise to some ingenious innovations, with designs for bicycles and benches also in the mix.

A 2021 report in Euronews demonstrates yet more potential for these blades, which is just as well, considering that vast quantities are on the cusp of reaching the end of their planned lifespan. These quantities are quoted as reaching 15,000 by 2023 in the UK and EU alone.[33] Potential for retired blades in Ireland include again pedestrian bridges, reinforcing components in the rail sector, portacabins and pods and even temporary access roads. Outdoor furniture, electrical transmission towers, highway noise barriers and coastal wake breaks are also under investigation.

Pyrolysis is another possibility when it comes to recycling. This involves first chopping the blades then breaking up the composite in a controlled process at high heat. Recovered fibres can then be used elsewhere, such as reinforcing concrete and using char residue as a fertiliser. It is however an energy-intensive process.

It may be interesting to mention that natural fibres are being investigated for use in making up these composite blades. Fibres such as jute and sisal have been investigated, bamboo too. In fact, bamboo is showing some promise in this application, but problems arise beyond those technical difficulties published such as quality consistency and wicking. Here again consideration must be made of these blades post-use. It is all well and good using natural materials in this application, but saturating a natural material within a solid shield of a polymer that can't be conventionally recycled may not be doing much to help the environmental cause. However, this is an area under research and perhaps this will pave the road to a future where we can produce energy using a truly green infrastructure.

10

TRANSPORT

Polymers in all their guises are so essential in all forms of transport that it can be hard to know where to begin. It is certainly worth looking into the most significant features of the role polymers have played in advancing the automotive, aviation and marine sectors, making mass transport accessible to all.

Polymers are now central in advancing the drive to greener more environmentally friendly transport. Of course when talking of transport, material density plays a big part in selecting components – lighter vehicles results in lower fuel consumption – but its role in creating the transport of the future, a future less reliant on the internal combustion engine, must also be noted.

Of course, if we are considering green transport we must acknowledge the role the humble bicycle has taken from the onset. It is arguably the most popular form of conveyance in the world and is now a far cry from the bone-shakers that first traversed the by-ways. Once an amalgam of steel, rubber and leather, bicycles now can almost entirely rely on polymers in their production. From polyurethane saddles, nylon pedals, rubber tyres and high-tech carbon-reinforced plastic frames, so many elements of a bicycle can be manufactured from polymers. Of course this does not preclude other materials and for many bikes it is very possible to find varying quantities of steel, aluminium and even titanium depending on the price range and expected use of the bicycle. The bike helmet, that other essential feature of the modern cyclist, is one terrific asset that has improved safety enormously. These helmets are heavily if not totally reliant on plastics being most commonly manufactured from an expanded polystyrene (EPS) inner and polycarbonate (PC) shell. If you do have a PC helmet, please don't be tempted to paint or add stickers to its surface as this might undermine the integrity of the material.

Of course mention must also be made to all those other vastly improved safety aspects of modern transport. Strides taken in implementing and improving safety features have resulted in far fewer fatalities and serious accidents than encountered in the past. A transportation mode arguably making the most of these advances is the automotive sector, and so it is this that we next bring into focus.

Automotive

Polymers have allowed massive forward strides in transport, with F1 cars typically leading the way when it comes to advances in the automotive sector. Car designs have adapted to become more streamlined, materials have changed to reduce weight and engine efficiencies have seen marked improvements. Changes overall target one vital end point and that is fuel efficiency. This is important, whether that fuel comes in the form of traditional petrol or diesel, or from biogas or electricity. After all, a lighter vehicle means lower fuel consumption. No doubt a great deal is owed to advances in glass- and carbon-fibre-filled materials (a.k.a. fibre-reinforced plastics or FRP). Materials which have proven their capabilities in sport, especially high-octane sports, move into the commercial sector with great regularity. Cars now boast a greater proportion of their make-up, both interior and exterior, to polymers.

Changes have become yet more significant as materials become ever more sophisticated and designs alter to take account of the freedoms polymers can offer. Take the fuel tank for instance. Once manufactured from several welded sections of metal (steel or aluminium), the entire tank system was redesigned to be manufactured in one process. Now the tank is formed from one piece of moulded plastic. (This isn't so unlike making one massive milk bottle, as the polymer selected can be either HDPE or PP, and this in turn is processed via a blow moulding technique just as with milk bottles). Eliminating welds and corners removes potential weak spots and certainly helps with economies, and moreover, using a plastic moulded in one fluid piece goes a long way to assisting in fuel savings.

In fact, nowadays a great deal of the typical modern automobile is polymer-based. Depending on the statistic quoted, and of course whether figures are being quoted as weight or volume percentages, these levels are significant. According to some sources this can be anything up to half the volume of the total material usage in the modern car. As an aside, it really is interesting how these quantities are measured. According to the trade group, the American Chemistry Council, on average cars traversing the road today contain about 50% plastic by volume but only 10% plastics by weight, which does illustrate the differences in density between traditional materials, e.g. metal and glass, and plastics. This report also shares an American Department of Energy statistic which claims that for every 10% reduction in a vehicle's weight, a corresponding 6–8% increase in fuel economy results. This of course has positive repercussions on emissions. Even where cars are electric or even hybrid, this still is an important benefit. Less power needed to heft a vehicle into motion and maintain propulsion will save energy, whatever form that takes.

Another report, this time from Resource Innovations, presents the data a little differently again. Here the number of components making up a car is pulled into the equation. They number the parts used in a vehicle as 30,000. Of this number roughly one third are manufactured from plastic. What isn't clear is the size or integrity of these parts, but it is an interesting figure all the same. This report also numbers the plastics types used in the automotive sector and comes up with thirty-nine different types of polymers found within a car. It is safe to say that the vast majority of plastic car parts are made from one of the big four polymers found in the sector. These are the omnipresent and exceedingly useful polypropylene (PP), PVC, which arguably is the plastic of greatest global significance, both joined by the eminently useful polyurethane (last seen in the Construction chapter) and the tough polyamides (a.k.a. nylon). There is no doubt that plastics are instrumental in the greening of cars, certainly in terms of fuel efficiencies. Interestingly, talking of greening, research is looking to replace the nylon used in car production with a bio produced nylon – its origins are caster beans. Of course as we are now well aware, whatever the source material, this nylon, just like its petrochemical twin, is still nylon. Despite the bio origins, it will not biodegrade.

In additional to this greening (if not degradability), plastics are also instrumental in improving safety and this ought not to be overlooked. Safety is imparted by more than just the essential child's car seat (most likely boasting a PP shell) and polymer woven seat belt (commonly polyester, although the stretchier nylon has also been used). Safety has also been designed in through the clever use of materials. This can be anything from the energy-absorbing bumper to the crumple of a side panel.

Design and knowledge of material properties have gone a long way into making our car journeys as effortless and comfortable as we now enjoy. Gone are the days when an annoying rattle struck up when anything above a cruising speed was approached. The interior is less inclined to vibrate and hum as once was the case and your feet can now remain dry no matter how inclement the weather outside. The rattles of old, sometimes the result of the wrong material choice, sometimes the result of too many disparate parts, not united as they ought, or maybe incompatible when abutting, are thankfully being designed into obsolescence. The dashboard, once prone to set up a sustained hum north of a certain velocity (which in turn sets up a resonance), now remains in blissful silence. An understanding of the damping properties of materials utilised has seen this component manufactured from an array of acceptable possibilities. This could be the indispensable PP, or polycarbonate blends, or polyurethane. Each can reduce the vibrations that used to make some cars sound like they were launching for space.

Staying with dashboards for the moment, they like many components associated with vehicles have progressed enormously from their original shape and format. Originally doing little more than providing a backing panel for the steering wheel and ignition switch, they evolved to house so much of the instrumentation needed to keep the driver aware of the car's functioning and comfort. These dashboards can be stiff, covered or foamed.

Polypropylene remains a popular choice when it comes to dashboards and shows up yet again in the flooring helping keep our feet dry. Those fibrous mats at our feet, or even the carpeting within the foot well, is most likely to have been woven from spun PP fibres.

But maybe most worthy of mention is the bumper. The bumper has come a long way from the shiny chrome offerings that began its exit from automobile construction from the 1980s and onward. Now chrome bumpers are really only ever seen on vintage cars, while newer models sport increasingly sophisticated, and most likely, integrated incarnations. These bumpers have gone through several iterations with a number of plastics materials coming into and out of favour. Now several material choices are possible including the ever versatile filled PP, ABS and a blend of polycarbonate and PBT. Whatever the material chosen, the aim is to ensure that the bumper and indeed the front zone of the car will crumple on impact. This absorbs the impact energy and so aims to protect both the engine and occupants from the worst injuries resulting from a crash. These safety aspects continue into the side panels both front and rear. Clever design coupled with savvy material choices mean cars, although of course never free from inflicting harm, are certainly safer than they were, particularly on impact. And this includes impact with unprotected individuals such as a pedestrian or cyclist. One complication arising from the synergistic effects of associating the bumper so closely with the side panels means that damage to a bumper can mean replacing more than just that section should there be cause.

The continued move towards electric vehicles will enable a greater greening of the industry. Going electric can facilitate greater quantities of plastics entering parts of the car, yet conversely see a decline in other areas. While this might seem anomalous, consider the changes necessary in the core components in an electrical vehicle. An electric battery is in (and this can contain plastics), while the PP fuel tank is out. Indeed there is a degree of internal redesign in the working components of an electric vehicle with PP and PU used for a greater number of applications, many replacing the often more expensive and slightly heavier engineering plastics, as well of course as metals, or for that again being used in new areas such as brackets for the electric battery. Polycarbonate is a winner too as electrification of cars continues unabated. It is used a fair amount in sensors and these seem to beep from all corners and angles in new car models. Interiors also see a revamp and where a high sheen is required, the aesthetically pleasing shine imparted from those materials containing polystyrene sees ABS (acrylonitrile butadiene styrene) in increasing usage. Problematic areas within these cars remain associated with the battery.

So from the nylon strap handles above the doors that you can use to heave yourself from your PU foamed, PP stitched seats to the ABS or polyvinylidene fluoride emblem or nameplate gracing the front and rear of the vehicle to the acetal rail lifting the electric windows, polymers do a lot to make a car functional and lightweight. This is before we ever get to the rubber wipers, polycarbonate lights and thermoplastic polyolefin (TPO) airbag chute. Through all this is a move to a circular economy when it comes to the automotive sector. Design is key, with an acknowledgement that this must ensure longevity, recyclability, and disassembly. The ability to recover all these plastic components post-use and to actually be able to do something with them other than add to landfill is an essential.

Before moving away from the automotive sector, it is worth having a look at tyres. Often forgotten except where a service or MOT is needed, tyres of course play a fundamental role in the efficiency of the vehicle. Even driving on a tyre at an incorrect pressure will lead to unnecessary wastage of fuel. Tyres, depending on their destined use can be moulded predominantly from natural rubber (now mostly found bouncing an aircraft to a halt on a runway or speeding a lorry along a motorway) or from a synthetic rubber called SBR (Styrene Butadiene Rubber) for cars. For most other vehicles a combination can also be found. In fact, there are very many ingredients involved in making tyres. Indeed ingredients is an apt word in this case as making tyres is not too dissimilar to mixing up a cake. The core rubber component in itself needs very many additions to obtain just the right properties required.

A very specific recipe needs to be followed where each ingredient is dosed in specific quantities at specific times. Ultimately each incorporated component will perform a very particular job. Once the rubber compound has been made, it in turn needs to be layered up onto other elements within the tyre including steel wires and nylon cords. The difficulty recyclers face when dealing with notoriously challenging tyres emanates partly from the rubber crosslinking when heated and partly to the myriad of elements constructing the tyre.

Other than recovering calorific value following incineration, many tyres find a second life crumbed and blended into slightly pliable outdoor surfaces such as those found in children's playgrounds, running tracks or similar. These crumbs can also find their way back into a new tyre when it can act as another filler ingredient in the tyre mix. To give an idea of the diverse and manifold elements making up a tyre, according to a research document by WRAP,[34] an all-season passenger tyre includes thirty kinds of synthetic rubber, eight kinds of natural rubber, eight kinds of carbon black, steel cord for belts encircling the tyre, polyester and nylon fibre , steel bead wire, up to forty different chemicals, waxes, oils, pigments, zinc oxide, silica and clays. Each have their own job to do, some such as carbon black and clay are fillers, the wires and fibres build in bands or plies and provide reinforcement, the oils act as plasticisers, sulphur vulcanises (or crosslinks), while additional chemicals provide anti-aging, or anti-oxidation protection.

Tyres are made in several stages with layers built up, formed and cured. They really are complex constructs with each layer very precisely put together. Continental Tyres give a concise rundown showing how a tyre builds from the road surface to the tyre inner. They list out these layers together with the source material as; tread – (natural and synthetic rubber); jointless cap plies (nylon, embedded in rubber); steel cord for belt plies (high-strength steel cords); textile cord ply (rubberised rayon or polyester); inner liner (butyl rubber); side wall (natural rubber); bead reinforcement (nylon or aramid); bead apex (synthetic rubber); bead core (steel wire embedded in rubber). All in all these unprepossessing lumps of rubber are highly complex and cleverly engineered. They efficiently and effectively keep us on the road and deal with multiple weather conditions.

On the road

The automobile has been given due consideration, but what about the roadways on which these vehicles travel? They too contain a surprising amount of polymers. We've already noted in an earlier chapter that road markings are more than just a slap of paint with formulations including polymers like epoxy or acrylic. In fact these formulations tend to be mostly polymer with some pigment added. Road lines are noted as we drive along but not necessarily noticed as an entity in itself. There are a lot of them, particularly in urban areas. They can be great dashed lines along the middle of the road, doubled on the edges to restrict parking, zig-zagged to notify motorists of an upcoming crossing and what of the crossing markings themselves. There is in all an awful lot of paint used.

Road safety features abound these days, but this was far from the case in the early days of motoring, and in the UK one of the earliest examples is the Belisha beacon, named after the transport minister who introduced them in the 1930s. They comprise yellow globes fixed to black and white striped poles with studs on the road to mark crossings. The early globes were manufactured from glass, but as plastic materials evolved in the early 1950s these beacons transitioned into globes of illuminated polyethylene and it is this that we are now likely to find dotted about our roads.

Many other initiatives came in the wake of the beacon and proliferated with the assent of polymers which made many easy to implement. One such are the cats' eyes set into the surface of the road. These were a clever invention by Percy Shaw in 1933. Here we have another example of a cat supplying the touch paper for an invention. When driving at night Shaw used the reflective gleam of tram lines embedded into the road as a guide keeping him on the correct side of the road. One particular night he found himself struggling to keep to his own side in a road section where the lines had been removed. Driving on cautiously he noticed that his headlights picked up a double pinpoint gleam of light. These transpired to belong to a cat sitting on a fence and serendipitously facing the right direction. This set up a chain of thought that evolved into the cat's eyes found embedded into roads throughout the country. These have aided countless motorists on dark nights with scant visibility. They are both clever and simple in their function with differing materials used in harmony.

They comprise metal and glass discs set into rubber and held on a metal base. The base is set into the road in such a way that rainwater drains into it while the rubber, inset with the reflective disc or bead, sits up proud of the road surface. When a car drives over the glass 'eyes' the rubber compresses, pushes down into the base, displaces residual rainwater, which in turn squirts up past the discs and so cleans their reflective surface before they pop back up again. The rubber too can act like a wiper to clean these surfaces. The system is pretty much self-sustaining with the reflective surfaces able to function without constant maintenance or cleaning.

Other road safety measures abound on the roads. There is much by way of street furniture certainly and much of this is either dominated or subsidised by plastics. These are vast in terms of range and scope. Bollards, signage and a host of traffic calming methods all contain plastics to varying degrees. A glance down any urban road will be rewarded by many examples within a very short distance. Bollards for instance may have started their roadside career as repurposed cannon but they have come a long way since. An early form of recycling saw redundant upturned cannon used as bollards in East London following a victory at the Battle of Trafalgar in 1805. These cannon, stripped from defeated French ships, didn't fit the English fleet so to cock a snoot at the French they were used as bollards on London streets. Some originals can still be seen but on the whole more modern urban versions are now used.

Those traffic bollards appearing at crossings, traffic islands, roundabouts and dotted about other locations on the road are now sometimes square white and yellow illuminated blocks of plastic. Many meet their doom at the hands of careless drivers and their smashed remains sprawl across roads so now they increasingly are found as more two dimensional plastic uprights that can be driven over with impunity, ready to spring back upright once crossed. These are manufactured from polyurethane or the ever popular HDPE or LDPE depending on their area of application. Even the roadside signage, which on first glance appears to comprise of painted metal only may be deceptive. Certainly the metal dominates, but there is a good chance the paint will contain polymer elements particularly if it is either an acrylic or alkyd enamel.

Cones dot the landscape, not so much street furniture as temporary interlopers. Uniform lines march along miles of roadway; they are a well-known sight to motorists across the land, and even make an odd guest appearance in student bedrooms. Bright orange or sometimes yellow, they are usually met with a sigh of resignation or frustration, they are there to slow our journeys. Stout and brightly coloured, these great lumps of plastic can be manhandled easily enough for quick deployment and if struck they offer good impact resistance. Plastics such as polypropylene

(PP), PVC or polyethylene (either LDPE or HDPE can be used) can all be used for this application being easily coloured, of low density, resistant to hydrocarbons coming from vehicle exhausts and able to survive inclement weather and even the odd knock from a car.

Aviation

Transport of course includes other modes of traveling such as aviation. There is no doubt that polymers have made their presence felt here and in a very positive way environmentally. This is most notably reflected in fuel consumption, where every ounce of weight wishing to become airborne results in greater quantities of fuel burnt. Even seemingly small changes such as the switch from glass to plastic miniatures on the drinks trolley has led to huge savings across a fleet.

This substitution has been so successful that other saving potentials are continually being assessed by airlines. Some of these are remarkable. Probably one of the most publicised is the so called $40,000 olive. In 1987 American Airlines decided to review its offerings to passengers to see where savings could be made. The simple action of removing a single olive from their in-flight salads not only made a saving of $40,000 across the year, both in fuel and grocery bills, but their passengers didn't even notice the reduction.

Not to be outdone in 2008 Northwest Airlines made an extraordinary saving of $500,000 a year by the simple expedient of slicing limes for in-flight drinks into sixteen pieces rather than the ten pieces previously used. Again much of this saving was seen in fuel reduction as well as savings in citrus purchases. These fine slicings are to be found across many airline fleets with successful measures including rationing in-flight magazines, or at the very least making them thinner, no longer offering paper receipts for purchases, cutting the quantities of ice carried, and cutting the need for so many paper manuals in the cockpit when a simple iPad would suffice.

Here we may meet with a moral dilemma. Using disposable cutlery on an aircraft will certainly aid in saving fuel but now we need to address the issues posed by using single-use plastic. How do we mitigate its use as opposed to reusable metal (heavier) cutlery? The need for a sensible workable balance is much needed – after all, why save oil by not using disposable cutlery only to burn it in the extra fuel needed to keep this metal in the air? In this case maybe it's a matter of pushing for more adept recycling. We'll come back to this issue in a moment. In all implementing greater instances where polymers can be incorporated in an aircraft will certainly go a long way to cut fuel use and therefore, of course, emissions.

While the skin of commercial aircraft remains resolutely metallic, with aluminium the material of choice, there is much research in this area. Dramatic changes to the fabric of an aircraft will not come quickly, as safety plays an enormous role in these matters, and any potential alternatives will be painstakingly tested. Research is considering fibre reinforced plastics (FRP) as widespread alternatives, but these are far more sophisticated than the FRP found on a for instance a car side panel. Layering fibres will give terrific strength along the length of those fibres, but it won't be so good transversely, where fibres can just be pulled apart. This is solved by laying a second layer of fibres perpendicular to the first – or, better still, weaving those fibres into a grid. This improves the strength in two directions, but what if tension comes in skew-whiff? Pull at a bit of fabric and see what happens, does it yield on the bias? Of course you can keep laying up fibres in all sorts of angles but then you end up with an overly complicated and maybe even an overly thick section. This is where knitting might offer a solution. Have a look at a knitted jumper, or better still have a go at knitting a line or two. The fibre twists, turns and interlocks in a 3D pattern, now there is a good distribution of fibres facing the right way at any given time to take up tension across a multiple of angles. Where that fibre complies with safety requirements, use is made of polyester or epoxy coated carbon, glass or aramid fibre, or indeed other newly developed materials. The aim is to offer good strength and stiffness, and if you have ever noticed how an aeroplane wing might flex in-flight, be reassured that good fatigue resistance is also an essential requirement.

While knitted aeroplanes may still be on the drawing board, composites can play a part in aircraft manufacture. According to the trade association, Composites UK, Airbus uses FRP in several areas including the fin, doors, undercarriage and tail plane. They quote a 20% replacement of aluminium alloy on the airframe which they say offers a weight-saving of 800kg over its equivalent for the Airbus 320. This of course will have major repercussions on fuel savings over the life of the aircraft. This all just considers the exterior fabric of an aeroplane. Once we get inside there is far more plastic in view. In the 1950s scant use was made of plastic, now a great proportion of what we can, and indeed can't, see is polymer-based. Speciality high-performance plastics such as the mouthful that is polyetheretherketone (PEEK) or polychlorotrifluoroethylene (PCTFE) are used in the internal working of an aeroplane in, for instance, pumps and valves where both high and low temperature resistance is key as is fire resistance, superior mechanical properties and low density. Maybe more visible to the passenger is the carpet (wool, nylon, or a mix of both), seatbelt webbing (nylon or polyester), the internal windows (PC or acrylic) or the seating shells. The seats actually can be built up with a number of material types. The frames may be aluminium or a carbon composite structure while arm rests could be polycarbonate (PC) or a blend of PVC and acrylic. Seat cushions may be a foamed rubber or polyurethane while the seat cover may be an easy to wipe down luxurious leather (or fake leather), wool, polyester or a blend. A number of polymer types can be used in a cabin interior. From the humble PP to the more sophisticated ABS and PVC blends, plastics can be found in many parts of the aeroplane from the wall shells and overhead panels to those items we hope never to see, the hidden life vest beneath our seat or overhead oxygen mask. The aim is to provide light-weight bright materials than are functional, possess the requisite mechanical and chemical properties, can be fire resistant and, now more importantly than ever, easily and hygienically cleaned.

This just considers those integral parts of the aeroplane. We started this section looking at expensive olives and considering disposable cutlery. There is no doubt that in the name of hygiene airlines use prodigious amounts of plastic in the guise of packaging, this of course is single-use and soon becomes waste. Swathes of plastics overwrap everything from the aforementioned disposable cutlery, pillows, blankets, earphones and even napkins, single toothpicks and those little salt and pepper sachets. Everyone, especially now, understands the need for hygiene but there must be middle ground. So much waste is generated on each flight yet much is unnecessary. Some airlines are very aware of the issue and are taking steps to change their practices but on the whole this problem remains. To add a layer of complexity to the issue, the very nature of an aeroplane being to move from one country to another can set up its own problems. Certain international flights mean touching down in countries with very tight biosecurity laws and impose tight restrictions on foodstuffs entering their borders. Not all countries allows the recycling of waste that had proximity to foodstuff from another country because of the risk of disease, particularly animal disease.

CHAPTER 10

Trains

Passenger trains have had a tendency to lag behind air transport when it comes to incorporating plastics in their structure. Of course both benefit from weight reductions, although air transport does see a greater correlation with fuel saving. Both also must consider safety aspects when considering materials used in construction with impact, smoke and fire resistance of high importance. Aesthetics too play their part, particularly internally.

Trains have come a long way since their great hulking solidity embodied those first steam trains. Iron yielded to steel and aluminium and now this is joined by fibre reinforced plastics (FRP). According to the dedicated news site Aluminium Insider, steel and aluminium remain the dominant materials in train locomotive and carriage manufacture with the sideboards, roof, floor panels and cant rails formed from these materials. With aluminium having a third of the weight of steel, increasing its proportion in construction will have an impact on energy consumption. This can be further improved with the introduction of carbon fibre materials in the form of FRP. Other FRP can be used in in the manufacture of the carriage body or shell with glass and aramid fibre also popular. These can be bound with polyester or epoxy with phenolic also present.

Materials used in conjunction with rail transport, particularly polymer-based materials, tend to be specially developed for their application and are designed now with an eye to sustainability. They also need to cope with the demands imposed on them during their service life. London's Underground stock for instance has to be particularly adherent to very strict fire regulations, a strictness that precludes the use of many materials on its cars. FRP development pays particular attention to this factor in addition to complying with strict standards on smoke and toxic fume suppression. In all they are finding increasing use particularly in refurbishments where they replace metal and even asbestos.

Internal panels, grab rails, seating and flooring now all can be, and are, produced from polymers of varying form. Interesting and incredibly strong foams and honeycombed sandwich structures can surround the carriage from the flooring to the ceiling via the doors. Those can be produced from a combination of materials both metal and polymer, and because their structure incorporates a great deal of air, they can be very lightweight. Fitting out the carriage also has an eye to weight-saving, so whether you are sitting on a regional train seat that bears more than a passing resemblance to an upright ironing board or luxuriating in first class intercity comfort then your seat probably has a great deal of plastic in its manufacture. These seats are not unlike those in the aviation industry in terms of structural materials, even down to the little drop down table incorporated into the seat back. Like the aviation seats, these too can be covered in a range of materials from high class leather to nylon, polyester or wool. Padding will come via polyurethane, silicone or polyester-based foams.

So when you lower your posterior onto the cushioned seating of an intercity train or stride across the foamed sandwich panels of the flooring (and not get heels caught in the wooden slats particularly prevalent in the Tube trains of yore) you can be thankful that a great deal of the comfort experienced emanates from polymers. The train driver can also be reassured that all possible is being done to improve the impact resistance of the cab. Again the energy-absorbing capabilities and panel construction methods of the polymers employed go a long way to maintaining safety. The cab can also benefit from a greater streamlining with fewer individual components requiring assembly and more incorporation into single components.

Of course it has to be acknowledged that where thermosets are used, and they are certainly used in this application (but not exclusively so), consideration needs to be made of sustainability. The sector understands this, but maybe needs to demonstrate how it is implemented. Scrapping of metal post-use is an understandable process: crush the metal, melt, reprocess. It might be beneficial to the sector if there was greater clarity with the post-usage of all, including polymer materials. As in other sectors, research is underway in developing natural fibre and bioplastic reinforcements as alternate sustainable materials. They aren't quite there yet, but it is an area to watch.

On the water

Boats have come a long way from their formative years many millennia ago. Evidence from prehistory has shown boats made of animal hides stretched over a wooden frame, pitch- or bitumen-coated woven reeds or hollowed-out tree trunks. These over time grew in size and complexity until we reach our current situation where the seas and inland waters are dotted with craft ranging from great hulking steel ferries and cargo ships to one-man dinghies and paddle boards.

While metal may dominate those larger craft, polymers can still of course be found; however, it is to those smaller vessels that we'll turn our attention. This is where plastics have made the greater impact both socio-economically and in terms of craft design. When the central aisle of certain supermarkets now offer a plethora of paraphernalia such as wetsuits, water-shoes, life preservers and even basic craft, such as paddleboats and kayaks, then affordability is within the reach of so many more than ever before.

Wetsuits are, by and large, produced from polychloroprene, a.k.a. Neoprene. Wriggling into these suits has increased in frequency with the accessibility of low-cost means of getting onto the water. Paddleboards for instance can come in a range of quality and prices. At the cheaper end would be those that are inflatable. Much more robust than a child's inflatable ring, these inflatables tend to consist of a double-skinned bladder of PVC. Manufacturing materials of choice can then extend through a broad range of prices and offerings from coated fibreglass (using epoxy or vinyl ester), glass fibre or wood laminated over an expanded polystyrene core, to the higher-end carbon fibre and epoxy constructions. Kayaks and canoes are also a common sight on many rivers with polyethylene (PE) being the material of choice for kayaks. Economical and low in density, PE can be thermoformed in two parts and sealed together, or rotationally moulded in much the same way as a council salt bin. These too can be easy enough to access. Likewise with canoes which can hit the water in a variety of guises.

Wooden canoes are still available; no longer hollowed-out logs, these tend to be carefully (and expensively) crafted from ash, birch or other locally available wood. Aluminium became the material of choice post-war but this now is pretty much superseded by lighter and easier to shape polymers. A range of materials can be selected depending on the requirements of the craft. Cheap and light polyethylene canoes can be manufactured like kayaks using a rotational moulding technique. Powdered PE is loaded into heated steel moulds and slowly rotated as it heats and eventually melts. As it is swirls about the inner cavity, the PE evenly coats the mould to the desired thickness. The mould then cools and out pops a canoe ready for further prepping before launching on the water. This technique is also perfect for those larger hollow items used on the water such as buoys, pontoons, dinghies and floats. Moving up a level come those canoes fabricated from a composite material. Depending on requirements this can be glass, carbon or even polyaramid (Kevlar) bound with epoxy, vinyl ester or polyester, sometimes also incorporating foams or even aluminium spars. This makes them lightweight and easy to repair.

Fast and versatile, inflatable boats are made possible through polymers. They can be found as recreational vessels but can also be seen zipping up and down water courses with a motor fitted into the end. These latter boats make for very effective safety or rescue boats, whether patrolling a stretch of water or working behind the scenes in regatta or sailing events. These inflatables come in two distinct types, either rigid or soft. The rigid inflatables tend to be more robust, whereas the soft ones lend themselves to deflating and transporting easily. They can be produced from rubber materials such as polychloroprene (like the wetsuit), PU or PVC which can in turn be coated onto other materials such as nylon or polyester. In addition the rigid variety can comprise a hull manufactured from aluminium or fibre-reinforced plastics (FRP) with an inflatable skirt.

Sailing boats can also rely quite heavily on polymers for their fabrication. The vessels themselves can range from tiny one-man dinghies to larger yachts, sloops or catamarans. The common feature across each is dominance of polymers, with glass fibre again present in much that appears on the water. Predictably carbon and polyaramid can provide more expensive alternatives, particularly where those craft are to be used in international sporting endeavours. On the whole, a variety of material choice is out there depending on the would-be sailors' budget and preferences. Wood of course is still present as a hull material, as is PE and metal. However FRP constructions, sometimes containing a foamed core, tend to predominate.

Sails and ropes are also usually polymer while the masts are wood, aluminium or high-end carbon fibre. Sails have come a long way from the previous favourite cotton, the material of choice for over a hundred years. Now top-flight sporting sails can also be manufactured from Kevlar, its trademark golden hue probably an added incentive to the sailors skimming the water beneath. Otherwise, polyester is a common material of choice. It could well be that some of the sails spotted on the water may well have at one time contained water in that serendipitous link sometimes provided by recycling. After all, many an old PET water bottle can end up as polyester fabric. In fact this material might as easily turn up in swimwear, surfing kites or windsurfing sails. Ultra-high molecular weight polyethylene (UHMWPE) can also be used, as can nylon, although this is only used on the smaller sails as it has a tendency to stretch. Ropes, sheets and halyards can be produced from many of the same materials with polyester and nylon commonly found along with polypropylene and UHMWPE (this is a far more specialist form of PE than that found in a plastic bag or milk bottle).

As touched upon, other items associated with water and travelling on the water can find their bases in polymers, and this includes PE buoys and pontoons and the now essential life preserver. According to the RNLI, life jackets were first invented in 1854 by a Captain Ward who developed a cork based jacket. This early jacket comprised long lengths of cork, a bit like oversized wine bottle corks, sewn in parallel onto a canvas jacket. This combination proved far more efficient than the tested alternatives which included balsa wood, rushes and even horsehair. Nowadays, these jackets tend to comprise an outer fabric skin of polyester or nylon (sometimes coated with thermoplastic polyurethane for added waterproofing) with an inner foamed material. This can be a foamed PE or PVC, whose job is to provide lightweight buoyancy and effectively achieved by trapping air within foam.

The larger life rings commonly found on river banks or sea fronts were once made from canvas covered cork. They now are produced from an HDPE filled with foamed polyurethane (PU).

These are attached to polypropylene ropes which allow the ring to be tossed to the person in need and then drawn back to the shore.

In all, plastic is far better seen used on the water as it ought be where it can offer so many freedoms and provide much enjoyment. Where plastics appear in water as flotsam and jetsam this is not acceptable. Its presence needs to be addressed as does a means of ensuring it does not appear there uninvited. We will examine just this issue a little later.

11

GOOD TO BE GREEN?

So far we have uncovered the various nooks and crannies where plastics can be found in in our lives and considered ways to reduce usage. Perhaps its prevalence, as uncovered in these last few chapters, might be surprising – or then again maybe not. Some might even say that 'shocking' would be a better choice of word. Yet this book was never about glorifying nor yet vilifying the material. Rather the aim was to give a balanced opinion.

Plastics, whether we like it or not, are pretty much everywhere. Modern society would find it difficult, if not impossible, to do without them. The abstraction of polymers in all their guises from our life would leave a very different landscape to the one we now enjoy. We would be forced to employ other materials in their stead, and while some may welcome the switch, it is questionable whether the swaps would be in any way beneficial to the planet in environmental terms. It is not always clear just what is involved in getting various materials into circulation. We are told how bad plastics are environmentally, but is this really the case when we take a step back and look at the entire picture? We are very aware how careless society has been in terms of its disposal, but what about actually getting plastics, and indeed polymers in general, into circulation? How do polymers compare with other materials in this respect? How sustainable are the alternatives? If we want to avoid plastics, how would we make those goods we need or have come to depend upon? What materials would we use? Where we use metals, how are the ores extracted and processed to give us a workable material? What about paper and textiles and glass and so many other materials?

If we were to ban plastics, or utterly curtail their use, would we in fact still want the products they are used to make? If so, how would we make these? All materials go through a production process and none are particularly kind to the planet. We would need vast quantities of alternative materials including natural materials. A lot of questions have been raised here, so now let's take an overview of the processes required to make some of these alternative materials, as well of course as plastics. One common factor across the board relates to the vast quantities of energy required to bring raw materials to a desired form. We really shouldn't forget the impact the production of this energy has on the environment, with much still generated from burning oil and gas. Nuclear is also in the mix (with its own challenge of waste disposal) and biomass an option increasingly adopted. Again, we do need to question the source of this biomass. Certainly some is waste produce, but some includes peat, wood, straw and other harvested raw material grown specifically to burn for fuel. Much of this has in its lifetime acted as a carbon store, carbon that is released on burning. We could use solar or wind, but the equipment necessary to capture this energy is in itself reliant on plastics, as we have touched upon already.

To say the issue is complex is putting it mildly. Part of our problem lies with the metrics used for proving or disproving the environmental credentials of a material. Where a carbon foot-print is quoted, a great deal of provisos needs to be put in place. The basic formation of certain materials (like concrete) emits CO_2 as they are created. This is a straightforward measurable chemical reaction. Here CO_2 is a tangible and direct by-product of the manufacturing process and this is the starting point for any calculations. Other materials react in other ways and may not specifically pump out CO_2 at all, so their carbon footprint can only be measured via their consumption of other emitters such as energy. But what energy? If a tree is transported from Norway to a paper mill in Scotland, do we assume the lorry ran on diesel not biofuel? Maybe we want to make the stats look good, so we input data indicating road transportation solely relied on biofuel vehicles, or that it arrived via an electrified railway. What about the sea crossing? What routes are calculated? It might seem insignificant, but it really does make a difference. The tree needs pulping and boiling to make paper. This requires electricity. How is *that* electricity produced? Does it come from a wind turbine on the mill roof or from the national grid, where it can as easily have been generated from fossil fuels?

Like the energy suppliers that advertise their electrical supply as being somehow greener and better than their rivals, this can be misleading. The supply entering your home comes from the national grid, just like anyone else purchasing their supply. It is not possible to extract electricity created from different starting points. It is just electricity and all fed into the one system no matter what the source. So again, how do you measure or even discern the carbon footprint of a material that might have been made on a day where the wind turbines were up and running on full speed, contributing their max to the grid, or on a fallow day when the wind wasn't quite right. The best we can do is to take an average, the worst we can do is look at the best- or worst-case scenario depending on what agenda is forwarded. This is part of the reason why carbon footprints can throw up diverse figures. Unless there is some clarity in the figures presented it might be best to just stick to the basics. It is a conundrum and not easily answered, so it is useful to have an understanding of the alternatives available to us and make our own minds up from there.

Let's start with concrete.

Concrete

Concrete is so ubiquitous in the built environment that hardly a second thought is directed its way. Grey and brutal, it is easy to overlook, yet it provides the very basis of so many construction enterprises. We tread on it daily as we traverse concourses on our way to those buildings that provide housing or workspace. It is versatile and can be poured, cast or formed into blocks. And what about its little brother, cement? Cement is the very glue holding our homes together and a constituent part of concrete.

Concrete has been in use for many thousands of years, with early iterations being used in the Middle East a good three thousand years ago. Tribes inhabiting the region provide the first evidence of an understanding both of the material and its chemistry. Several other civilisations used mortar in their buildings, but this was not strictly speaking cement as we would define it since they did not base this glue on limestone (the key component of cement). For instance the Great Wall of China uses a mortar to hold the building blocks of this edifice in place, but this mortar was based on a gloopy sticky rice, a surprisingly robust material. So although early Middle Eastern tribes were known to use a form of concrete, it was employed on a small scale before being later used in earnest by the Romans.

The Roman formulation was deployed across the empire then lost in the midst of time. It was glimpsed briefly over succeeding centuries then resurrected at the tail end of the 1700s, when it was used by John Smeaton in the rebuilding of the Eddystone Lighthouse. Smeaton called this cement forbearer hydraulic lime, as he discovered that it would set under water, a very useful attribute when you are tasked with building a lighthouse.

Cement then took on a new life in 1824 when Portland cement entered the arena. The Portland moniker came about when Joseph Aspdin revised Smeaton's cement and noted that it looked a little like the Portland stone then used for building, particularly where a high-quality stone was needed. Aspdin was a bricklayer by trade and dabbling with various concoctions in his kitchen (which must have made him popular in his household) he hit on a very effective method for making cement.

This cement was formed by grinding quantities of limestone and clay and burning in a kiln until carbon dioxide was released. It is this later requirement that we will come back to in a bit. The temperatures needed were enormous as the raw materials had to be heated to such an extent that they fused together. The resultant fused lumpy 'clinker' could then be ground. At this point Aspdin had his cement. This cement eventually found its way into modern concrete.

Concrete consists of cement (to around 10–15%), aggregate (usually stone and sand) and water. Nowadays additional components are added to tailor the concrete according to use. These are chemicals that will provide resistance to a number of external elements under different scenarios and does so by controlling the way in which the concrete sets. This can provide quite nuanced concretes that can stand up to a sliding scale of temperatures and even inclement weather, including high winds, all depending on the recipes used.

Now cement is vital in so many facets of our lives (in concrete or as mortar bonding bricks), whether building the next energy substation or the newest skyscraper. Cement production is a huge global industry, with China, perhaps unsurprisingly, providing by far the world's greatest market for the material. Modern methods of producing cement may have started in the UK, but it is a material success story throughout the world. Environmentally, however, that success is somewhat tainted, as cement production does have a huge impact on the planet. This is acknowledged by producers, and steps are underway to green the process with some success. However, those newer forms of cement production that are kinder environmentally go hand in glove with higher production costs, and so not all construction projects are willing to adopt these changes and it remains quite niche.

Environmental issues are firmly rooted in the processing of cement (and therefore concrete) with two aspects particularly troublesome. Processing temperatures are exceedingly high needing to hit 1,300 to 1,450°C. To put this into context, molten lava reaches temperatures of about 700° to 1,250°C. Needless to say, reaching such scorching temperatures necessitates vast quantities of energy. Basically, like the volcano, the cement kiln is looking to change the state of rock, only this time not to liquefy it but to take it one step further.

The equipment used in cement production is enormous, taking on board great quantities of limestone and clay, which is heated and slowly rotated as it passes along a rotary kiln until the point where the desired sintering reaction takes place. The aim is to break down the elemental components of limestone separating out carbon dioxide (CO_2) from calcium oxide. Limestone itself ($CaCO_3$), just considering its calcinated composition, comprises something like 40% CO_2; the rest is the calcium oxide (CaO). It is the calcium oxide (now in the form of a clinker) that is needed for the next step. This clinker is cooled, fed into a mill where it is ground to a powder then mixed with other powdered components, including gypsum, to form the final cement. The CO_2 is not needed and of course is of great environmental concern not least because of the sheer quantities produced. It is estimated that for every ton of calcium oxide converted, between 622kg and 750kg of CO_2 is released, depending on what report is quoted. This is the CO_2 released by the raw materials on conversion to cement. When considering in terms of carbon footprint, which varies with raw materials used and methods of energy production it is widely recognised that something like 7–8% of the global CO_2 created comes from the manufacture of cement. To put it into context, this is far greater than the 2.5% used in the aviation industry. The only sectors producing more CO_2 are transport (taken collectively) and energy generation. Of course the industry understands where problems lie and have proposed solutions including carbon capture, but it is the development of other forms of cement that will really have the desired impact. In the meantime, although cement and concrete is not in frequent competition with polymers, it does rival it in certain domestic applications. For instance concrete has now become a trendy urban décor choice used for flooring, exposed walls, poured worktops and of course structurally as support beams. Each of these uses has polymer alternatives, as do other competitors, including wood or metals.

Concrete is one of the most widely used substances on earth (according to a BBC article, it is the most widely used to all but water), yet reaching back into history arguably metal took this honour. Metal in all its guises was certainly used for many important aspects of life whether the gold band twisted to signify marriage, the sword used to possess or protect land or the ploughshare used to cultivate that land. Metals require all manner of processing both to mine and extract ore. The furnace temperatures, and energy required to convert metal are perhaps widely acknowledged but maybe less known is the sheer quantity and extent of chemicals and/or processes needed in the extraction and conversion process. The cocktail of chemicals used remains in great quantities of waste water and lies in great dammed lakes known as tailings. One such recently broke in Brazil with disastrous consequences.

Several steps are required in producing metals and these differ according to the metal extracted. Here we'll take a closer look some.

Copper

Copper and sulphide are closely associated in a mining context as the majority of copper ore mined is in fact a copper sulphide. Copper deposits occur in varying concentrations and purity levels geographically but wherever sourced it does tend to be surrounded a fair amount of waste material called gangue. To separate this unnecessary rock, the mined ore is ground to a fine powder then passed through a floatation process where the gangue settles on the bottom and removed. The powdered copper sulphide is then heated to a temperature high enough to trigger the chemical reaction needed to instigate the process of removing the sulphur. The temperatures used aren't high enough to melt the ore (called calcine at this point), at 500°C to 700°C it just starts the reaction and dries the ore after its watery floatation.

Next a substance is added to aid in the melting of the ore. The calcine is heated to 1,200°C a temperature high enough (and more) to liquefy it. Other impurities such as iron sulphide can be present still and at this stage are removed. To further purify, the liquid copper is cast into anodes and then goes through an electrolysis process at which point it gains a high degree of purity. Between the heating and electrolysis, this is a pretty energy-intensive path to copper formation and now an alternative method also comes into play. This second method involves leaching the ore to extract the copper. This is a method using far less energy but unfortunately necessitates the use of a dilute sulphuric acid. Dilute or not, sulphuric acid is an unpleasant chemical. The leaching entails pouring the acid over the ore allowing it to seep though over a period of time (months rather than hours). The acid picks out the copper sulphate and holds it in solution ready for the next stage of electrolysis. As with the high temperature method (smelting), this brings the ore into a high form of purity.

It is not unusual when recovering metals to have a choice in the methods of recovery. Take gold for instance. It can be found in veins large enough to extract nuggets or in finer veins within quartz crushed for recovery. The finer particles of gold can be extracted through heating to temperatures sufficient to melt the metal or through a heap leach process that involves diluting cyanide in water and allowing it to drip through crushed gold bearing rock or gold rich soil. The gold, now in solution, passes through carbon particles which capture the gold as it flows or is agitated through. The gold rich carbon is collected then burnt off leaving the gold. The waste water remains infused with cyanide.

Another method involves forming the gold into an amalgam with mercury. The mercury, which is highly toxic, is burnt off leaving the gold behind. Gold has always been a status metal used sparingly and for specific applications, electrical as well as for adornment.

Steel

Like concrete, steel is everywhere in the built environment and more. It can be found in precision medical blades or the robust shafts of hip replacements (not to mention in the billions of hypodermic syringe needles used globally, particularly now). It can be found in kitchen drawers and cupboards up and down the country forming cutlery, utensils, pans, appliances and more. It's within reinforced concrete and on almost every major form of transport we wish to use. It has widespread usage, yet we barely give it a second thought.

Steel has been around a while, but it is the inventor Henry Bessemer that turned the corner in devising a cheap effective means of production and so making it available as never before. It wasn't long before steel found its way into ships, the new railway tracks and the steam engines of the day as well as (eventually) skyscrapers. Bessemer's father had fled the French Revolution, having worked for a while as a mechanical engineer making machinery for the Paris mint. He settled in Hertfordshire before moving the family to London, where the young Bessemer began his career as an inventor particularly in metallurgy. During the Crimean War he devised a processing method for making strong steel for gun barrels. It had mixed success. However, he went on to establish a steelworks and refine his process which involved deploying a specially designed vessel allowing blasts of air to travel through the molten metal, trap and remove carbon impurities and strengthen the metal. The process over time tweaked out the issues that had made prior output unreliable, and in time the Bessemer method became a global standard in steel manufacture laying the foundation for the cheaper more efficient production methods used today. Creating steel relies on a couple of integral core steps.

The process involves combining iron with coke. This is produced from coal, which, is fed into an oven heated to around 1,100°C and left burning for eighteen hours. This carbonises the coal bringing it down to its core element and removes impurities such as coal tar and coal gas. The coke that remains is then cooled and joins iron ore and limestone, which in turn is sintered and fed into a blast furnace. Here air heated to 1,000°C is injected into the bottom of the furnace via nozzles and as the coke burns, temperatures reach 2,000°C or more. This creates molten iron, which sinks to the bottom of the furnace while the limestone binds with impurities to form slag, which floats. The slag is removed and sent for use in the cement and road building industries.

To convert the molten iron to steel another process is needed and here high purity oxygen is blown onto the surface. This oxygen bombards the molten iron at twice the speed of sound using very high pressure. The resultant steel can then go through additional processes depending on the grade and quality of steel required. A quantity of carbon remains within the raw steel and is controllable so for instance a structural steel benefits from higher levels of residual carbon as this yields greater strength. Once optimised as desired, the steel is drawn, cooled and cut.

Steel-making is environmentally a dirty process and is incredibly carbon-intensive. It is blamed for around 5% of EU and 7% of global CO_2 emissions, partly because of the coal used within the process and partly because of the incredible levels of energy needed. A great deal of this energy comes from coal-hungry power plants. Of course alternatives are available, but at the moment steel is produced in great volumes in parts of the world that currently favour coal-fuelled energy.

Aluminium

Aluminium is another material imbued with a degree of environmental impact not commonly recognised. Production of this metal begins with extraction of bauxite ore from the earth. Bauxite is the ore usually needed for aluminium production and according to the group European Aluminium; much is mined in tropical locations, in part due to its high energy requirements. Energy does tend to be cheaper to produce in these areas. Looking at statistical output levels, Australia, Guinea and China are among the three most dominant bauxite producing countries.

Ore is mined from just beneath the earth's surface, and is extracted by removing surface soil and mining the rock beneath. This is ground to smaller particle sizes then refined by treatment with a concentrated sodium hydroxide (aka caustic soda or lye) solution under elevated temperature and pressure. The resultant solution (which has become aluminium hydroxide) is filtered, precipitated then heated to a temperature of around 1200°C. This calcinates the solution to give aluminium oxide. Now a smelting process removes the oxygen atoms within the aluminium oxide leaving just the pure metal. This involves electrolysis, an energy-intensive process but one most effective at breaking the strong bonds binding the oxygen to the aluminium.

The process entails putting the aluminium oxide into solution, placing in a carbon-lined vessel, inserting carbon anodes and passing huge currents through the solution. The whole process runs at elevated temperatures of up to 1,000°C. The carbon anodes are consumed during the process and many scientific papers pinpoint this consumption, combined with the emission of oxygen atoms from the solution, to result in the formation of CO_2 and CO (carbon monoxide). The pure aluminium metal forms on the bottom or the vessel from where it is syphoned away, cast into ingots and/or alloyed with other strengthening metals.

Environmentally, pollution is created at various stages in the process, with gases given off along the way including carbon dioxide, carbon monoxide and fluorine. This is before the vast quantities of electricity needed for the process are even considered, to say nothing of the aforementioned sludge that needs damming and to be kept from water courses. The sludge effluent from the aluminium process is known as red mud and contains all the impurities extracted from the solution as well as the chemicals used in the process. The volume of this decanted red mud is vast and highly toxic with millions of tons produced annually, so research continues to neutralise this output. The difficulty remains in making any potential solutions scalable and safe. For any budding chemist out there, this is a global problem in need of an effective solution.

The counterargument saying that aluminium is endlessly recyclable is, I believe, not strictly true, and either way should not deflect from its troublesome production. The pure material may indeed be infinitely recycled, but life isn't always that easy and chances are there will be some element of contaminant, no matter how small, fed in at each recycling stage. It all means that we do not have a closed loop whereby aluminium going into new products comes directly from recycled metal. Leaving aside our present inability to collect all redundant aluminium, new and expanding markets mean there will always be a need for yet more production from bauxite, and it is here that the greater issues lie in terms of environmental impact.

Using water packaging as an example where aluminium cans seek to replace plastic bottles, a 2019 report by Reuters examined just this issue. Taking account of production, recycling levels and even how energy used in production was created (hydro power versus fossil fuels), they found each can contributes around twice the carbon output of an equivalent plastic bottle (for Europe). Taking the worst-case scenario in terms of production, this report found a 330ml aluminium can resulted in 1,300 grams of CO_2 emissions while a plastic (PET) 330ml bottle was responsible for up to 330 grams. Of course this is the worst-case scenario and clearly stated as such in the report. Taking a kinder view towards aluminium and allocating it a greater quantity of recycled material within the mix, and assuming energy has been derived solely from a renewable source, the CO_2 emissions can be brought down to a little under 200grams per can. The likelihood of all energy being renewable is at this point of time unlikely, and so a figure somewhere in the middle is more reasonable.

Right now the route to production, together with its consumption of elevated levels of energy makes it difficult for aluminium to present itself as a better choice environmentally. It really is a very complex issue and not at all as clear-cut as many in the industry would like to present. Certainly when it comes to canning water, I really would advocate considering whether the product is really needed before plucking from the shelves or indeed heeding the glossy advertising.

Glass

Glass is an old material around for millennia, and is not without its own issues. Originally natural forms of glassy materials such as obsidian (a naturally occurring volcanic glass) and rock crystal were carved into shape by early users. When exactly the jump to manufacturing to order came about is not known for sure. Archaeologists have uncovered man-made Mesopotamian glass that dates back 5,000 years, although it's possible to push that back another 2,000 years to the Phoenicians. Either way, man-made glass has been around a very long time. For much of its early existence, glass-making was a time-consuming process with the high temperatures needed to melt the material proving difficult to master. The breakthrough in this proto-production came with the Syrians a little over 2,000 years ago when the concept of the blowpipe was introduced to the process.

This manufacturing aid made a great difference to glass production and really sped up the process. Then the ever-resourceful Romans entered the arena and began to produce ornate glass goods and utilised the power and extent of their empire to spread their process techniques far and wide. By the Middle Ages many centres of excellence had sprung up with glass used in many ornate and functional applications.

The next major development in glass-making came about in the 1670s when the Englishman George Ravenscroft devised an improved technique of making lead glass. Not much is known about Ravenscroft other than his employment as a trader importing and exporting all sorts of commodities including currents. How he came to glass-making isn't so clear, but many sources do attribute this form of glass to him. Crystal glass was lauded for its brilliance and clarity, making it very appealing to users of the time.

Another breakthrough in the glass making process came about relatively recently in the 1950s, when Sir Alastair Pilkington devised a method of producing flat glass via a floatation technique. So effective is this technique that even now the vast majority of plate glass produced uses the Pilkington method. Alastair Pilkington was a mechanical scientist charged with making distortion-free quality glass. This was important for any glass destined for use in mirrors for instance. At that time flat even glass needed polishing because of the roller marks made during the manufacturing process. This was expensive. Sheet glass was cheaper and formed by drawing directly from the furnace, but this also distorted and was not of a high enough quality for many domestic applications. The answer came from floating the molten glass on a flat surface.

How this is achieved is quite remarkable. The molten glass is drawn in a continuous line upwards and outwards from the furnace then floated on the surface of molten tin. Drawing across molten tin means that the glass is on a flat surface, is kept at very high temperatures and allowed to settle flat and blemish free. As the glass floats along its bath of tin it is gradually cooled to a point where it can be removed without rollers marking the glass surface. It took a long time to develop this entire process, but, as mentioned, it now is a method employed globally.

We may have wandered a little here looking at processes before considering the constituent parts of glass and its basic formation. A number of raw materials are needed for glass – more than just the familiar sand. These can include sodium oxide, calcium oxide, aluminium oxide, potash, boron and feldspar. Lead can also be used when making crystal but luckily the radioactive thorium oxide is no longer used. Of the sand (or quartz) component, silica is favoured as it has a high melting temperature. This is in the region of 1,500–1,700°C, just enough to give a treacly viscous melt. Some ingredients added into the mix aid the viscosity of the molten glass making it easier to form. Others are present to differentiate between glass that is used for tableware (fancy or plain), bottles and windows, glass that will happily sit in an oven or lab bench, or even glass that has high optical qualities that find use in cables.

The whole process requires very high temperatures, as mentioned, which will melt the constituent parts and cause the requisite chemical reactions. These reactions in part include the formation of CO_2 which vents away as the chemical reactions reach their conclusion. The UK government published a report in 2019 which in part looked at the CO_2 generation during the glass-making process. Approximately 360,000 tonnes of CO_2 were calculated as the yearly output from the industry. This isn't the carbon footprint; it is the direct creation of CO_2. As mentioned at the start of this chapter, measurement of carbon footprint is fraught with difficulties. This figure relates to the tangible production of CO_2 caused by chemical reactions during manufacture (the total CO_2 generated per annum was cited as two million tonnes taking energy requirements into account). The key to positive environmental change here, as in so many industries, is the increased use of recycled glass (known as cullet). Adding cullet to the process will allow the scrap to be reformed. The environmental advantages of this cullet lies in the fact that any chemical reactions will have taken place when first formed and so no additional CO_2 needs to be emitted. Recycling glass does make a big difference. CO_2 is not the only gas formed during the process. Depending on the base ingredients used, oxygen, sulphur dioxide and other gases can be released.

Glass production does of course have environmental impacts. Much of this sits with the very high temperatures needed to maintain the production furnaces. Here as elsewhere, alternative sources of power are being considered, with hydrogen probably offering a workable alternative to the natural gas normally used. It is not a solution applicable tomorrow, but it will hopefully one day make a real difference to this energy-intensive process.

Burning gas does, of course, have very real environmental consequences so any effective alternative would be welcomed. It is not an area commonly acknowledged in environmental terms; instead we do have a tendency to focus on the worst aspects of plastic. Of course it would be utterly wrong to claim that plastics offer a wondrous form of salvation. Naturally they bring their own baggage with them and it is up to the individual to decide whether any of this is mitigated when considered as a whole and comparatively to alternative materials. So it is to plastics that we turn next.

Plastics

All material processes have an impact environmentally and elsewhere, and plastics are obviously no exception. So how do plastics come into being? Plastics, and particularly polymers, can come from a number of sources including natural, but it is those that derive from oil that have remained firmly in the spotlight, so it is here that we initially turn our attention. It has been mentioned several times already that plastics are a subsidiary outcome from the oil and gas industry, with the main purpose of extraction being to deliver fuel. However this balance may change in the future. A report published in March 2023 by Natural Gas World predicted that 2050 will see a doubling of oil derived plastic production. They expect the growing demand for plastics to offset reductions otherwise expected as society transition away from fossil fuels in other areas. Whatever the future predictions, currently just 4 to 6% of extracted oil and gas is used for plastics production, Of course it is only fair to state just how this extraction and processing takes place before we shine the spotlight on plastics specifically.

Oil and gas extraction, whether on land or sea, involves drilling and pumping. (Oil being the more complex process, we'll continue examining this substance.) In the early day of oil extraction bitumen was washed with hot water to separate the oil from sand and clay. In fact in certain instances this method can still be employed, particularly where the oil is contained within surface sands and can't be pumped conventionally. Like pretty much all surface mining, waste water is held in tailing ponds. In this instance no chemicals are needed for the process, and so the tailings consist of just water, sand, clay and some small oil residue. The sand and clay settle in time and the water is reused for the process.

The majority of extraction operations of course involve drilling with gas deposits generally found with oil. The oil can come in various levels of viscosity depending on the position it sits within the earth. There is light, medium, heavy or extra heavy oil with light or medium oils capable of flowing naturally. Generally it is these grades that are drilled and pumped in offshore operations, as the oil can flow under atmospheric conditions and is the most economical to recover and subsequently process. Although it is an uncomplicated extraction, there is still a need for heavy equipment and enormous rigs and platforms to be established when this extraction takes place at sea. The rigs are generally built on land before being towed offshore. There, depending on the size of the rig and how deep the waters, they are secured in place and drilling commences. Securing a rig can involve piles, mooring or anchor systems. Whatever method is used, the rig must remain stable through high seas and aggressive winds. This drilling format is a common method of extraction. There are other more costly recovery methods, both in terms of expense and energy consumption but as these are classed as unconventional extractions they are not commonly found.

Crude oil contains hydrocarbons, and hydrocarbons mean energy and this energy needs extraction. On average the components of crude oil can be broken down as being 84% carbon, 14% hydrogen and up to 3% sulphur. So once this crude oil is recovered it needs refinement. Depending on the type of crude oil extracted (light to heavy), it undergoes a series of refinements and upgrades to extract residues or reduce its sulphur component. A process called fractional distillation separates the oil into fractions by passing it through a tall bottom heated column. As each of the components (that is the hydrocarbon as defined by its molecular length) within the oil has differing boiling points, each will condense at a different position along the column. In this way the heavier bitumen (which is denser with a longer molecular chain length) for instance can be separated from the lighter kerosene (having a shorter molecule).

A process called cracking further breaks up the large complex hydrocarbon molecules. By heating to around 700°C (which is enough to vaporise the molecules) these large molecules break down into much smaller, more manageable hydrocarbon molecules. Alternatively the hydrocarbon can be passed over a catalyst of silica or alumina, which will give a similar outcome. Basically, it is these smaller molecules that are needed to produce plastics. Another process involves hydro-treating which stabilises the oil and removes sulphur and other impurities such as metals or anything that will interfere or cause damage in any engine or machinery it may end up in. This treatment passes the oil through many stages under heat (up to 350°C) and moderate pressure while reacting with hydrogen.

Hydrocarbons, as their name indicates, consist of carbon and hydrogen. For plastics, it is the smaller alkene (carbon-to-carbon) element that needs extraction particularly when forming the building blocks of those familiar plastics, polyethylene (PE) and polypropylene (PP). A method called steam cracking can be employed for the next stage, which entails breaking the hydrocarbon into its constituent parts, so providing the starting point for PE and PP. Here the hydrocarbon is combined with steam at elevated temperatures, over 800°C, and high pressure. Once broken down, the smaller ethylene or propylene molecules need to be built back up in order to produce polyethylene, polypropylene or a number of other plastic types. Various catalysts can come into play here. This is where it does get a little difficult to give an overarching account of just how all this takes place, as there are so many different plastics, many derived from differing chemical blocks or elements such as chlorine, oxygen or nitrogen. Where these are bonded tightly together, a greater quantity of energy is needed to separate them, so allowing them to join up into the blocks needed for a particular plastic.

Because of the complex nature of plastics it really is difficult to give exact energy consumption figures, not least because much of the data is quoted in joules and so not necessarily comparative across material types. However, when you are looking at energy needed to break bonds it does mean you are going to need quite a lot of it. To add to the confusing picture, some plastics are exothermic, or in other words they give out heat energy as they form. There are plans afoot to bring this back into the loop during production. In general, however, the actual processing of thermoplastic materials (those that can be heat-recycled) necessitates heating to around 180–300°C, depending on the plastic processed , to form into the desired shape.

Plastic's environmental impact is somewhat complicated and entangled with all those other possible products deriving from oil and gas. Of course it is now possible to produce plastics from plant sources, but how is this achieved?

Bioplastics

Plastics derived from biological material such as plant, fungi or even algae or yeast, are known as bioplastics. These bio-based polymers can be manufactured to give widely ranging material properties. We have already noted that the 'bio' element of the term can be misleading, as it does not always indicate biodegradable, rather the source material has bio origins. These can be synthesised in a number of ways with some bioplastics created by extracting polysaccharides from crops such as maize (a.k.a. corn), tapioca or sugar cane. Polylactic acid (PLA) is probably the best known bioplastic of this type with maize a popular raw material. The raw material is first ground in a mill where starch is extracted (starch being a polysaccharide). This is then combined with an acid or enzymes and heated. The starch now converts into a substance called dextrose or corn sugar, which in turn is fermented to yield lactic acid. The lactic acid does present some problems in getting it to join into the long molecular chains needed to make PLA. It gets there by a couple of chemical reactions that first makes short-chain PLA before these are joined again to lengthen their molecules.

There is an alternative route to production and that involves fermented bacteria. These very specific bacteria can be broken down and reformed as a biodegradable bioplastic called polyhydroxyalkanoate (or, less of a mouthful, PHA). They are useful as they degrade in a number of environments including soil and marine environments. Where waste materials such as potato or banana peel and possibly even whey can be used, these processes have even greater appeal. This may be one of those materials to watch, but right now the route to production is very costly.

There are a number of moral debates to navigate here too. When there is hunger in the world, how many crops ought we divert to plastics manufacture rather than address the hunger? Of course there are many bio sources that would not make for good food-stuff, but looking at the current situation, productive crops are just that and can be used for either end point. To further muddy the picture, some academic studies comparing bottles derived from fossil fuels with those coming from bio sources point to the oil-based bottles being best environmentally. This in part is due to growth factors including planting, fertilising, applying pesticides, harvesting and so on. Water usage too plays a part as does the discharged irrigation water, sometimes laden with fertilisers and other chemicals, has on water courses. Where more complex polymers such as PET are being more closely replicated, the issues become yet more complicated.

Synthesis of certain bio-based plastics can be very energy-intensive (many are roughly comparable to existing plastics) and therefore, considering their role in environmental terms, do have pitfalls. PEF for instance, which is so like PET that it can be recycled in the same waste stream, is so energy-intensive to create that it was shelved when first produced. It was easier and cheaper to stick with PET. This material is now being revived because it ticks certain boxes. PE can also be derived from bio sources (sugarcane for instance) and will have green credentials emblazoned across its products (a lot in the packaging sector). No matter what the origins, PE is PE and can't biodegrade. We do need to be careful of demanding bio materials just because they are not oil-based only to risk causing more environmental harm. A range of bio materials have the potential to create bioplastics when their atomic building blocks contain the right elements. Cleaving these apart is not easy though and on the whole takes a lot of energy with more energy needed to build them back up in the desired pattern. Green origins do not necessarily equate to green outcomes. However, research does continue on a range of possible sources to satisfy our requirements across a range of applications. These include using CO_2 itself as a source material in the production of plastics. Of course, as mentioned, this does entail much energy use in order to split apart the molecule to reach the carbon atom. Much work remains to make these alternatives scalable but there is hope in viable alternatives to existing options. A lot depends on workable energy sources. Where energy can be produced cleanly and efficiently, so much more could be open to us.

Paper

The next material to enter the spotlight is paper, and here we have hopes of a clean sheet in terms of environmental concerns. Sadly we are to be disappointed in this expectation, although it must be said, in terms of comparative energy consumption and direct CO_2 production it does fair better than many alternatives. Starting with the very basics, paper requires the felling of trees. Those that are felled release quantities of the carbon they have sequestered, depending on the way they are processed. Replanting is all well and good, but full-grown trees are substituted with saplings which do not have the carbon-capture abilities of their elders; however, it is renewable, and efforts are being made to ramp up forestry. It is worth trying to improve tree populations. A typical tree can absorb in the region of 10kg (22lb) of CO_2 per year. However, this figure has many provisos to consider, including tree variety, where it is located and its maturity.

What of the paper manufacturing process? Here we have a material that actually does grow on trees, but to get it from the tree and turned into something useful is a little more involved. First you need to select the right trees. Those chosen tend be a mix of soft and hard woods with combinations offering just the desired properties. Of the softwood trees, species selected include spruce, fir and pine, whereas the hardwood trees preferred include eucalyptus or birch. Get the balance right and the resultant paper can encapsulate anything from soft, pliable toilet roll to top-quality bond paper (the later can also incorporate rag mixed with wood pulp).

Whatever the ultimate destination, the sheer quantity of trees needed to sustain the current paper industry is substantial. It is estimated that in the region of four billion trees are cut globally each and every year just to give us paper, and this is a conservative average, some commentators cite a much higher figure. Modern means of communication really should have reduced the amounts of paper we use, and maybe in some sectors this is the case. But of course many industries are also turning back to paper for packaging, as many chose to shun plastic, and let's not forget toilet paper. The empty shelves in the early days of the pandemic maybe showed us just how reliant we are on this particular product! Whatever the reasons, we go through reams of the stuff nowadays. According to the CPI (Confederation of Paper Industries) there are forty-seven paper mills in the UK alone. Although they churn out around four million tons of paper per year, we still demand more and consume ten million tons a year (This figure being the most recent one available according to a report by Statista Research Department in February 2023). The CPI point out that the UK imports more paper than any other country in the world. Is so much paper usage a good or bad thing? Maybe we can decide once we understand how we get paper.

The paper-making process of course starts with the tree, which is stripped of its bark and chipped or cut into logs. The bark has other uses, maybe as mulch or maybe as a biofuel. The chips (or shortened logs) can now take one of two routes, either of which involves parting wood fibre from lignin (the binder holding cellulose fibres together) and pulping. As a tree consists of up to 30% lignin, there is quite a bit to disentangle or even remove.

One route follows a mechanical path, which is an efficient method of creating the fibre, giving workable yields anywhere between 90–96%. There are a couple of options here too. In one lengths of debarked logs are torn apart by huge steel discs like a giant joint of pulled pork. The second option involves grinding the logs between grindstones and pulping the wood in that way. Where chips are processed these can be pre-softened by steam treating to make them easier to grind.

This mechanical method just pulls apart the lignin and doesn't actually remove it. The test of time will tell if paper production has taken this route. Lignin is light-sensitive and degrades over time, giving a yellowed appearance. For this reason mechanical pulping will lead to the production of cheaper paper and paper that is destined for books or newsprint. This is the paper that discolours and weakens over time (partly because grinding will give shorter fibre lengths). Recycled paper (deinked and pulped) takes this route too. The pulp itself will be darker and requires bleaching before it can be used in paper form. Of course, if great lengths of tree are to be ground, this means a great deal of energy will need to be expended. Just how much energy does depend on a number of factors, not least on the wood type itself with a hard-wood species requiring more energy than soft wood. Water too is needed and at this stage in the process it is sprayed onto the grinders to keep them cool while they work.

An alternate route to pulping involves a chemical process. Here the wood chips are heated (up to 130°C) and put under pressure in a chemical environment. The lignin is loosened from the fibre by chemicals such as caustic soda and sulphur. The process is not as efficient as the mechanical route, which yields more fibre. However, the fibres are longer and therefore stronger and the resultant paper is of a higher quality and not so likely to yellow with age. It is a high-energy process. Again, as with the mechanically pulped fibre, bleaching agents ensure resultant paper can be white or lightly coloured.

Next the pulp is suspended in water, up to 99 parts water to 1 part pulp in order to get a very dilute thin fibre solution. The pulp is screened to remove impurities and at this stage can also be mixed with pulped recycled paper. Once cleaned up, the highly diluted pulped fibre solution flows onto a plastic mesh where, drained of water, a thin layer of matted fibres is left. The paper is carried on the mesh while water drains away until the conveyor carries the paper to the next phase. Now the paper is drawn through mangle-like rollers which squeeze out more water. At this point a thin layer of starch can be laid on the surface of the paper to stiffen it – this is known as sizing. Then it passes through a series of heated drying cylinders where it further dries.

Next the paper can go through additional stages. It can be further sized or coated with other chemicals, pigments or substances such as china clay, chalk or even latex, to impart a variety of finishes. The paper can also pass through rollers with rough or polished hard or soft (or both) surfaces depending on end requirements. For instance, magazine paper requires a thin compressed paper with a high sheen. This is accomplished at the roller stage.

In all the process does use enormous quantities of water, especially at the dilution stage. The pulped fibre of each tree makes up just 1% of the solution with water being the remaining 99% before being meshed. That is a lot of water. Of course there is an awareness of the impact this has environmentally, so there are steps to make this more sustainable. What can be recovered and fed back into the system is, but difficulties do lie here, particularly when considering the quantities of chemicals that are infused into the water. These can be difficult to remove and so does make recycling or even discharging the water a challenge. Great strides have been taken over the past few decades to reduce untreated effluent from entering the environment. The intensive use of energy is also a recognised challenge to the industry. Pounding a tree to pulp takes a lot of energy, as does maintaining an elevated water temperature. In all it is recognised that a lot of resources are consumed in this industry and a lot of waste, both solid and liquid, needs to be treated. It is an on-going issue.

Textiles

Textiles cover quite a broad range, encapsulating those made of both synthetic and natural fibres and indeed a mix of both. Production of synthetic fibres overlaps with information given already (in the plastics section), so the focus here will be directed towards natural fibres. In addition to that information given on plastics production, much publicity surrounds polyester fibres spun from ex-PET bottles. This is a very useful method of recycling, with PET reappearing as wadding, fleece, stuffing used for toys as well as reel spun fibres capable of being woven into a fabric. Producing fibre is a positive outcome for old bottles, although do remember to empty any residual liquid before tossing in the green bin. (Interestingly since the UK Plastics Packaging Tax was introduced in April 2022, recycled PET has become a valuable commodity, maybe too valuable for a new life as a fibre). Of course, even recycling clear bottles does not necessarily mean they can be used as a 100% base material. It may be remembered a few chapters back that each time a polymer gets recycled its long chained molecules gets chopped about. Shredding the bottles will reduce chain length, as will churning it through the processing equipment and squeezing it through forming dies. Each time the chain shortens, the resultant fibre will weaken, and so, to counter this, fresh virgin material is added.

Other fibre sources are available, such as acrylic, nylon and even elastomers. Again their route to production doesn't differ enormously from that given already, particularly if we are looking at energy consumption and other environmental impacts. So as we have a grasp on the impact of plastics let's look now at natural materials.

Wool

Not all wool is equal, with different breeds of sheep giving different wool types. This can span the range from very fine merino wool to much coarser crossbred wool. All have their place commercially with coarser wool best suited for heavy apparel and carpets, and finer wool for high-quality clothing. The properties determining where wool is slotted relates to its diameter and crimp. The crimp basically refers to how wavy the fibre is, and this depends on sheep breed.

Once you have ascertained the wool type and have shorn the sheep, the fibre can be processed. This entails a few basic steps. Processing begins with scouring, which, as implied, just means giving the wool a good wash. The process removes dirt and grease and leaves the wool ready for its conversion to yarn. Carding comes next, and is a similar process as found with many naturally occurring fibres. Carding teases out the jumbled mass of wool fibres, naturally entangled on the sheep as it grows but becomes further entangled following the scouring process. This step rakes through the fibres, unmalting them from their random tangle. Now the fibre is ready for gilling, a combing process which further aligns the fibres into parallel strands. Combing goes one step further in giving uniform long strands with any short fibres or unwanted matter removed at this point. Next comes drafting, a process that takes the fibres from a fairly thick strand down to a very fine strand. Now the fibres are ready for spinning and twisting into a yarn, then woven into a fabric, or indeed twisted into a carpet.

But what about the environmental impact of the industry? Certainly in the past it is well recorded (particularly in the highlands) that land clearances took place to graze sheep as the Middle Ages saw immense wealth enter the country though this sector. Now much of the global wool production takes place in New Zealand and Australia in the vast tracts of grassland found there. Sheep, like cattle, deer and antelope, are ruminants, meaning their stomachs are divided into four chambers, and this in turn means methane generation. However, there are mitigating points that work in favour of sheep. They graze from pastures which are in themselves carbon sinks and do not require much in the way of additional water (certainly not as much as cattle) and can get a great deal of their requirements from the grass itself. They can crop the grass closely but if they are allowed enough pasture they actually manage the land well. Have a look in a field grazed by sheep if you get the chance. Generally the biodiversity is good. They don't tend to allow rampant growth of any particular plant, allowing a selection to grow without being subdued. This in turn attracts a variety of insect and other life to these pastures.

According to the International Wool Textile Organisation (IWTO) there are other plus points for wool. A point not readily recognised is that sheep themselves are part of the natural carbon cycle. They consume the carbon stored within plants and in turn convert this to wool. The IMTO say that 50% of the weight of wool is in fact carbon. Added to this is the fact that wool has a long lifespan, tends to be washed less frequently and at lower temperatures. Wool is also recyclable and biodegradable.

Silk

With increasing numbers choosing to eschew all products having any taint of animal product in their making, it may be a surprise for many to learn how some products come to be. Silk is a good example. It is pretty well known that silk comes from cocoons woven by the silkworm. The worms (which are more accurately described as larvae or caterpillars) weave in one continuous strand building up layer upon layer of silk spun into a figure of eight as they encase themselves ready for their transition to a moth. This strand can be incredibly long, up to a mile long in some cases or sometimes even more. The silk strands are held together with a gummy type substance. If the larva is allowed to break from its cocoon, the silk thread is broken (an excretion from the larva dissolves the silk into short lengths easily broken through). Of course any breakages to that thread will devalue the silk and can make it too short to work with efficiently. A long strong continuous thread is the aim of cultivators, and in order to achieve this, the cocoon is boiled. Of course this does have the unhappy result of killing the larvae in situ. Boiling or steam treating kills the larvae, dissolves the gummy substance and loosens the silk enabling it to unravel and be reeled. The thread is then washed, bleached and dyed, then spun and woven.

In terms of energy or resource consumption, silk production needs the cultivation of vast quantities of mulberry leaves. According to the Council of Fashion Designers of America who produced an informative guide on the subject, in the region of one hundred silkworms can be sustained by a single mulberry tree. Yet 3,000 silkworms are required just to make one square yard of silk and this can necessitate chemical pesticides and fertilisers. Fertilisers, maybe counterintuitively, in the form of manure can cause a lot of harm to biodiversity, particularly where there is a runoff into water courses. Even so, not a great deal is used, certainly not in quantities approaching that of cotton, which we'll come onto next. Energy is needed at various stages in silk production from temperature control keeping the silkworms in a comfortable environment, to heating the water to dismantle the cocoons. Water too is needed in quantities sufficient to boil the cocoons and indeed to irrigate the mulberry trees. Water also comes into play for cleaning the silk. Here chemicals, some toxic, are added to aid the process. Synthetic dyes too can cause issues as they are cheaper and easier than natural dyes. Unfortunately, not all countries producing silk are on top of their waste treatment plans.

Silk is expensive and not a mainstream material, so let's turn our intention next to the far more prevalent cotton.

Cotton

Cotton has been around for a long time – how long exactly is not known with any certainty, as fabrics don't tend to endure archeologically. Remarkably though some evidence of cotton cloth has been found in Mexico and dated as being around 7,000 years old. Maybe even more remarkably, it appears that the cotton from that era is similar to that found today. This is unusual, as many cultivated crops have been adapted and tweaked over a great many years. Cotton was known across Mesoamerica but also crops up among other civilisations, including within the Indus basin and in Egypt. It continued being useful over the millennia, but generally, particularly in the western world, on a relatively small scale. Here we tended to stick with wool or, if we could afford it, silk.

Jump forward to the industrial revolution and all changed. Cotton had been steadily entering the country as the British Empire expanded. The East India Company operating in the region of the Indian Ocean traded in many commodities including cotton. Introducing it to the UK in larger quantities caused it to ramp up in popularity. Trade coming also from the West Indies increased its presence and popularity in the country. It was durable and easy to dye and wash. Then calico (a rougher not so processed form of cotton) was introduced, and now it came within the scope of even the poorest. Of course manufacturers within the country who processed from bales of raw cotton were not happy with bolts of cheap calico entering the country. A ban (the Calico Act) lasting two decades was implemented and again restricted the affordability of cotton.

Cotton was relatively expensive to manufacture, as the fibre needed to be spun before weaving. It was a labour-intensive manual process, with fibres formed individually. The industrial revolution, however, was a time of innovation, and it wasn't long before James Hargreaves devised the spinning jenny. This device could spin several fibres at once, greatly speeding up this process. The spinning jenny certainly revolutionised the fibre process but these fibres were found to be weaker than the handmade equivalent. Next Richard Arkwright entered the arena. He solved the problem with his invention of the water frame. This frame paved the way for large-scale production of now quality fabric. Britain soon came to dominate the global market, with the Midlands faring particularly well.

Cotton had a profound impact on Britain in more ways than one. It also had an impact on the countries exporting the raw cotton. Not all these impacts were good. Life was not good for the mill workers, nor was it good for those harvesting the cotton. Conditions and society may have improved dramatically from those days of revolution, but how about the environment? How sustainable is this product?

Today cotton is primarily produced in China, India, the United States and Brazil (in that order). Cotton can only be grown along certain latitudes and needs a great deal of water during its growth stage. Unfortunately the latitudes within which the cotton grows commercially is not always geographically blessed with an abundance of water or indeed rainfall, and this has caused problems in some areas where water depletion to feed cotton crops has led to serious shortages elsewhere. The problems begin before this, however. In order to plant, the land is prepped by clearing the soil of everything that might compete for nutrients and water or even attract pests. To ensure that the nutrient levels remain high and pests kept away, a great many fertilisers and pesticides (in the form of insecticides, fungicides, and herbicides) are employed. In this way yields can be increased as much as possible. The latter in particular is used in such high quantities that in America alone cultivating cotton necessitates the annual use of 25% of all chemical pesticides used on American crops. In addition to this, a chemical spray defoliates the cotton stalks prior to harvesting as it's easier to process the cotton if just the boll and stalk remain standing.

According to The World Counts (an organisation sourcing and condensing data from a large number of organisations, research institutions, news services etc.) cotton is the world's most chemically sprayed crop and accounts for 16% of global insecticide releases. It has been estimated that it takes 5.3oz (151g) of chemicals to produce one pound of processed cotton. Let's not forget that many of these are petrochemical in origin and so also come from the gas and oil industry. This is just the chemical input – billions of tons of water are also needed (calculated on average at ten gallons per cotton plant), as are high levels of energy to cultivate and process. It is worth noting that organic cotton has its own issues, and now a move is afoot (certainly in the US) to produce 'cleaner' cotton. Here the cotton will still have a chemical input, but these are curtailed where possible and more natural pesticides used. Not great for biodiversity, but at least this recognises the problems. Water use remains a problem however. UNESCO has gone further in their report by calculating that cotton consumption is responsible for 2.6% of the global water use. As so much cotton is exported from its country of origin, the report states that 44% of the water used for cotton growth and processing directly feeds overseas markets, and so as we increase our demand we are further reducing the water available domestically in those countries. Indeed their 2023 World Water Development Report states that agriculture versus urban demands on water projects the latter's demand to increase by 80% by 2050 so putting further pressures on this resource.

Transforming the cotton to fibre and then cloth sees it carded, which like wool basically combs the fibres into long parallel lengths, spinning into yarn or thread, then woven into fabric. Of course there is energy usage, here but not excessive, as no great heat or rigorous mechanical input is needed, although chemicals such as sodium hydroxide and hydrogen peroxide are used to wash and bleach the fibres. In fact sodium hydroxide is also used in connection with cotton and other fibres for a process called mercerisation. Sometimes seen as a cloth or thread descriptor, it just means that it has greater strength and can more easily be dyed. This leads us to the next environmental problem with the fabric and that comes with the dyeing process. It is important to say here that many fabrics, even synthetic, can have similar issues with the dyeing process. To dye fabric quantities of fresh pure water are required. The purity of the water is very important. Even an incorrect pH balance will affect the dye quality. Natural dyes can of course be used, but nowadays much is sourced synthetically (in other words very likely to be petrochemical). These dyes are dispersed in large heated baths of water in which the fabric is washed. Dyes, depending on how they are formulated, can give rise to a great deal of highly coloured wastewater. By rights this ought to be treated before returning to the environment. Unfortunately it is recognised that this does not always happen. Added to these issues, some dying techniques and formulations also call for large quantities of salt as a fixing agent. Again this makes wastewater treatment problematic, or indeed adds to the pollution issues where treatment is not carried out or waste unregulated.

Cotton is a complex commodity. Certainly there are issues with its production, but it undeniably is a very important crop. If steps can be taken to mitigate the problems in the sector it can actually do a lot of good. According to the United Nations, cotton is a life-changing product, so much so that they celebrate World Cotton Day in October each year. This is in recognition of the support it gives to the world's most impoverished regions. Indeed it also reflects the varied use the entire plant can have, from the seed to the lint and many other parts between. Its use as a textile is well established but parts of the plant can also provide animal feed, edible oils, cosmetics or fuel, among other uses. Also to its advantage, cotton does not require large tracts of land given over for cultivation, yet it supplies 27% of the world's textile needs. Maybe the key here is to address existing issues surrounding cultivation and production and not be so wasteful when it comes to this textile. It is also worth discovering the source point of the cotton – are producers compliant with various protocols issued with sustainability in mind? Some retailers are better than others when it comes to pressing for good practices amongst growers.

Leather

Leather has become somewhat an emotive subject, connected as it is to the slaughter of animals. Leather (and hide in general) is another of those materials reaching back in time. Its route to manufacture (known as tanning) has come some way (thankfully) from the urine-soaked hides processed historically. Tanning is the process of retaining the tough collagen within the skin while eliminating surplus fatty substances and hair. It has evolved in the last hundred years or so, with a number of chemicals now used during the process instead of urine. These can be chosen to impart a particular look, to be kinder environmentally (although this is still a major concern), or just on economic grounds. It may be no surprise that the latter choice commonly comes to the fore.

The process is complex, involving several steps, with hide preparation alone involving a cleaning soak, hair removal and liming (to preserve the grain of the skin), and bating to flush out the sodium sulphide and lime used in the previous step. The final step in just this initial stage of the process involves pickling. Here the skins are treated with sulphuric and/or formic acid to obtain the desired pH – this is important for optimal penetration of the tanning agent. Next comes salt to suppress swelling following the addition of the acid.

Tanning itself sees agents such as chromium tanning salts or vegetable tannins introduced. These are fixed chemically, then the leather is split, and shaved to a thickness dependent on the end use of the leather (for instance furniture requires thicker leather than clothes). Generally speaking the majority of tanned leather will involve the use of chromium. The tanning process can be repeated depending on the tone of the leather. And this is followed by dyeing, oiling (for softness) and drying (to fix everything in place). Additional layers of chemicals or oils can be added to the surface but basically the leather is now ready for use.

These are the basic steps taken to produce leather. Of course there is a lot more to consider if we are to take an environmental slant on the process. Leather has always been one of those industries sited at the very edges of a town, and for very good reason. Although the solutions used in the process may have changed, the high volumes of water needed persist, as does the often toxic chemical mix amalgamated within that water. According to the Central Leather Research Institute (CLRI) of India, in the region of 40 to 45 litres of water is needed per kilogramme of raw hide, and if you consider the weight of this hide, that equates to a lot of water. In fact rather than using the start weight, if we consider the weight of the finished leather, according to a more recent 2020 article by the Clothing Manufacturers UK, 15,000 litres of water goes to make 1kg of finished leather, with much of this water directed to the washing process. Whatever way you take the figures, it is a lot of water. This water can be reused within certain cycles of the process. For instance, water at the liming stage could be recycled for additional batches. This is not possible at each stage, as much of the effluent produced contains high concentrations of salts. This also makes it difficult to ecologically flush into the environment. Effluent from the leather industry is considered one of the most damaging ecologically.

Perhaps of greater concern is the level of chemical consumption during the tanning process. A 2021 scientific article appearing in the *Journal of Environmental Impact Assessment* quotes average chemical consumption to be 360.2 kg per ton of shaved leather. Taking this from another angle, something like 130 different chemicals are needed to produce leather when considering from cow to finished hide. Of these chromium generates a great deal of concern and is the subject of much work aiming, at the very least, to remove it from wastewater. As with many other industries, there is recognition of environmental concerns, with steps being taken to address these. Where these industries are based overseas, an extra layer of complexity is added, with practices not always easy to effectively monitor.

There are several opposing drivers in this sector both for and against including those that wish to grow this industry and those that wish to see an end to it on ethical grounds. Ethical arguments are for the individual consumer to consider. Our interested lies in the impact the industry has on the environment. Fake leather does of course exist and has done so for some time. These have tended to be of a synthetic variety mostly leaning on PVC or polyurethane (PU) in its production. Now, however, research is underway to create fake leather through a more natural process and that is to grow it. The process is still in pretty early stages, and so any material produced does take quite some time to manufacture anything sizable. No doubt this will improve in time. Mention has already been made of certain stores now marketing vegan products in the form of shoes, bags and garments. They do not contain leather, even though it looks as if they do. At this point in time most likely their origins will be in PVC or PU, yet they are marketed to appeal to a certain sector of society to generate more sales.

Coatings

Not a material in a conventional sense, varnish and other waterproofing methods are worth a brief look. An early original and natural varnish is shellac. It is very much still present on the high street, most noticeably in nail bars, but we do need to consider how effective, economical or even ethical is its production.

Shellac is an interesting material. It is produced by a small beetle (called a lac beetle) living on the Indian subcontinent. The beetle secretes a sticky substance protecting and securing its larvae onto tree branches. This sticky substance hardens into the substance known as shellac. It's harvested by lopping the branches, scrapping them clean and then heating the residue until it turns to a liquid. This liquid is filtered to remove any larvae, beetles or bits of bark caught during the scraping process. If it were just the residue that was processed then there not be a question mark over this process. Unfortunately, as with the silk-making process, live larvae and beetles are caught up in the wholesale harvesting of the branches and heated along with the rest of the substance. These are destined to perish in the process.

There are other waxes that cause less harm to their producers such as beeswax and lanolin (from sheep), and of course there are plant-based waxes. These include carnauba wax grown only from a particular palm tree in a particular region of Brazil, and linseed, which comes from the flax plant. Synthetic waxes are mostly hydrocarbon-based and are familiar as paraffin wax and can also be based on PE (polyethylene). These diverse materials turn up in all sorts of applications. Waxed coats for instance were first waterproofed by sailors fed up of getting drenched at sea. They used materials to hand such as fish oils. These were heavy and, unsurprisingly, smelly. Then linseed was found to be just as effective, although also quite heavy and aromatic! Now waxed jackets and other waxed clothing take advantage of neutral smelling, lightweight synthetic wax.

Again there is a balance to consider between growing the plants needed to create oils, and here palm probably has attracted the greater attention because of the destruction of native habitats in order to grow these trees. On the other hand there are those who do not wish to use any product derived from a living creature. And of course the concerns surrounding synthetic materials are well established. It is hard to find a route through this moral dilemma.

12

THEN AND NOW: SOME BETTER USES AND SOME HIDDEN USES

Mankind has always used the resources the earth has provided and manipulated these for our benefit and furtherance. Whatever age we are in, we exploit the resources available to the maximum, right back to the stone users of prehistory. Our ingenuity has always enabled us to circumvent difficulties encountered in life by developing new and better material alternatives. This has at times led to specific problems being shifted but not resolved, as will be seen as we delve deeper into some material uses.

On their discovery, metals were embraced for the many superior qualities they exhibited as compared to their predecessors. For instance we had copper, but it was soft, so we discovered alloys: mixing in tin resulted in bronze, and a whole new age was born. Soon we moved forward again into the Iron Age, where iron and then steel were developed. Jump forward a millennium and developments in cast iron enabled formidable edifices to be erected and vastly improved transportation methods effectively shrunk the world. Other materials moved us into increasingly advanced modes of living, with the plastics age propelling us faster and further than ever before. We broke from the confines of the earth and now reach for the stars. Just look at how much has been achieved in the last century, a century where plastics became mass market and indispensable to our way of living.

At each point in our timeline we have turned to materials to better our circumstances, always striving for a better understanding of the bounty of the earth and using it to best effect. This has not always been greatly advantageous to either our fellow creatures or the planet we share with them.

We tend to think of our present age as being mired with the ravages of pollutants. Were things really so much better years ago? Recent evidence of pollution created during the Industrial Revolution was discovered in ice core extracts from the Himalayas. Amongst other contaminants these extracts included heavy metals. Fortunately we now know to avoid or contain their usage where possible. Where such contaminants are emitted in the present day, scrubbers or filters are employed stop their emission into the atmosphere. This is the role a catalytic converter takes on in a car for instance. Similar convertors or scrubbers are used on an industrial scale in incinerator chimneys or can be found on a small scale to clean the emissions belching from domestic stoves. It would be foolish to think this happens everywhere or to think that Industrial Age pollution is in the past. It is only natural that emerging economies want to better the fortunes of their own people; unfortunately in many cases this entails large quantities of pollutants. This, of course, is a whole other issue; what we in this country need to consider is that industrial-level pollution has been going on for a great number of years and that the arrival of plastic was not the first cause of environmental blight.

It is a trait of human nature to look back at the past through a mist of nostalgia filtering out the worst aspects of life as it was. And so it is with those who long for the clock to turn back and have things just as they were in pre-plastic days. But being realistic, we should ask, is this possible and, more to the point, is it even desirable? Maybe to help the planet emerge extant through the myriad of problems it now faces, we should embrace the use of plastics. Needless to say this use needs to be sensible and balanced. But perhaps we ought to acknowledge the advantages plastics offer, advantages that perhaps extent to a greater degree than imagined.

With this in mind, let's consider whether it is sensible to turn the clock back and make sole use of the materials of the past? We can compare a number of products and consider their past manufacture with what is on offer today. One area under intense scrutiny relates to packaging.

Packaging

There is now a desire to see more paper in the shops for our packaging requirements. We have already seen how this is not always a practical alternative when it comes to pre-packaged items, but what about the paper bags that used to be bundled on loops of string and hung from a hook in the greengrocers? Originally these were replaced, first in the supermarket, then at the grocers, by those rolls of lightweight plastic bags that tend to be frustratingly difficult to open. Now these rolls of little bags are ebbing and flowing from sight, as supermarkets are prevaricating over their future role in our lives.

We have already looked into the weight discrepancies and thickness variations occurring between the two sets of bags. (Plastic possess roughly a 20% lower density than paper, and in terms of thickness, they can be produced to a much finer gauge of up to seven times thinner than that of paper. This means a greater number of plastic bags can be fitted into a given volume of paper bags.) We know that in terms of transport, plastic is less demanding on fuel. We are aware too of manufacturing processes and the steps involved in producing paper. And what about the plastics still used in these modern paper bags? Worryingly many do still incorporate plastic. Notice too that many are stamped with a certifiable wood source badge, but they do not always state any recyclable credentials. Any product containing a mix of core materials is going to make recycling complicated. Where these products are of low value it makes reprocessing less viable, both financially and in terms of energy and straightforward manpower required.

Plastic can be present even in a paper bag, and these tend to be those bags with a clear strip down the middle so the person on the till can see the produce without having to peep into the bag. This strip can be (but not always) made of cellulose. Fine, but what if it's heat-sealed onto the paper? Now it is highly possible that the cellulose is layered with a polyethylene strip which when heated creates the requisite seal. Remember: it's not just the seal points that are plastic-layered, it is the entire sheet. For the manufacturers this is the most cost-effective and efficient means for providing the seal.

Ideally this transparent strip would need to be torn away from the paper to recycle. Maybe some wouldn't mind doing this if the bags were of any use in the first place. I've yet to find one that will refrain from splitting if more than three apples are placed inside, and if put next to a damp product from the chiller cabinets then it is a near certainty the whole thing will spill its contents before getting near the till. So in terms of a positive PR exercise it works. The public want to see plastic banished. But when the replacement is not fit for purpose and still containing plastics, it is questionable whether this is a step in the right direction. And on one final note on this subject, it is interesting to see that an alternative option is tiptoeing onto the fruit and veg aisles and that is the sturdy reusable drawstring bag, purchasable for a small fee. I've made a couple myself from an old net curtain and bit of ribbon. They work, are certainly sturdy, reusable, and are made of plastic in their entirety. Interesting that the solution to one type of plastic bag should be another plastic bag. The point is, where it is used sensibly and consumers are offered practical reusable alternatives, it can work and help in the push to environmentally sensitive living.

There is no doubt that we are becoming more wasteful, though. As our lives ease and we have so much choice at our fingertips, it is very easy to forget the premium set on common materials such as paper. Perusing a book[35] written in 1940 aimed at advising the housewife of the day on a multitude of sundry matters, one finds directions for dealing with waste from the dust gathered by the daily floor sweep (burn it) to cans (bin them) to paper. It advocates not treating paper as refuse but instead to reuse it about the house or put it away for salvage. (As an aside, this book notes the increasing variety of plastics being made and advises that they are wiped over with a damp cloth and polished with a duster to keep clean. At this point there was little more available than Bakelite, or acrylic for dentures.)

We will revisit bags and the choices we have when it comes to how they are made in the next chapter. Before that, it might be interesting to have a closer look at some other everyday items. Each has been chosen for further scrutiny, either because they have an interesting history or they are used without much background knowledge, or because of the impact they have – particularly on the environment. In the examples that follow, plastics by no means are the panacea leaping in to make everything better. It will be clearly and firmly pointed out where they are causing undue problems.

Pot scrubs and other dishwashing paraphernalia

Now that we have some comparative information at our fingertips, why not look back on a more mundane element of daily life. Something we get on with mechanically without giving it a second thought, and that's doing the dishes. Of course there will be many households that give over this function in its entirety to the handy machine in the corner. Yet a pot scrub tends to be on stand-by and ready for action at most sinks. Whether it is destined for the post-dinner scrub or just to swill out an odd mug, it very plainly is a foamed plastic. Of course this little block of foam (in many cases it is a foamed polyurethane (PU) backed with a coarse layer of fibrous polyethylene (PE)) wasn't always available to us so it's interesting to see what happened before the humble pot scrub originated.

In the days before foamed PU and PE-backed pot-scouring sponges, not to mention cookware lined with PTFE (a.k.a. Teflon), it was a tough job getting everything clean. Pots left to soak could ease the job, but in the end it came down to elbow grease and the aids of the day. Sharp river sand was a common and well established method of scouring out stubborn encrustations. A quantity of sand and a good rub would eventually shift the most stubborn of burnt-in residue. Revisiting the Tudor era, a particular type of river reed was used with the sand, as it soaped up nicely and could cut through the greasiest of pan. So good was this reed that it was virtually wiped out in England and had to be imported from Holland.

If we were to banish the pot scrub, what could we use in its place? Do we go back to river sand and reeds? It caused an environmental problem in its day when the population of the country stood at roughly four million, how much worse now with a UK population approaching 67 million? Never mind the impact on habitat of removing river sand and reeds, what about the energy needed to excavate, package and transport heavy sand about the country? What would it do to the drains and how would we tackle silted sewers? Sand abrades, and pipes, certainly domestically, are increasingly plastic in origin (PVC or PEX, which is a type of cross-linked polyethylene). Abrasion of these pipes would most likely result in unwanted incremental levels of microplastics.

Maybe looking at Tudor habits is going too far back, although sand was still used relatively recently. The modern idea of the pot scrub came in the early 1900s with several versions patented or produced in the 1920s. Arguably it was an American inventor, Russell Kingman, who devised the scrub from which our modern version derived. Whoever is selected as the instigator of the scourer, in these early years a common factor entailed the deployment of a fine-wire wool. That's far too harsh on most modern cookware and flatware. It also needs to be manufactured. This means a starting point of mined ore, extraction processes and high energy consumption in melting and converting the steel (as detailed in the last chapter). Not to mention the fact that steel is close to seven times the density of the plastics otherwise eschewed and so is heavier, leading to the knock-on effect of greater energy consumption in transporting it.

All right, let's go back to dishcloths, not the modern disposable all-purpose type (made usually from a PET/viscose mix), rather the cotton ones that needed a good boil every few days (nowadays it is more likely a consumer will just replace these when they get grubby regardless of the environmental impact cotton imposes on the planet). Leaving aside the need for boiling to remove any harbouring bacteria, it could probably work on most cookware, although we're back to elbow grease for the stubborn stuff.

Cotton is a nice, natural material. It is not local though, and does need shipping. And we discovered in the last chapter, the processes involved in its manufacture can throw up their own problems. What about those 'green' replacement scouring sponges? Of course there are continued developments in providing an eco-friendly green alternative. It is possible to synthesise PU to ensure it will degrade post-use. Research continues, but at the moment its presence is limited and the route to degradation dependant on a number of factors. Even here, the main constituent parts of the polymer is petrochemical dependant, although here again research is looking into bio alternatives.

In a rethink of these scrubs, other older and more natural materials are being tasked with breaching the gap between function and green credentials. Here enters the humble loofah. Loofahs are increasingly creeping into this sector, but again it must be remembered they were widely discarded in the first place because of intrinsic problems, not least of which is its propensity to harbour bacteria and need for thorough cleaning themselves. With a large cellular structure, they can provide a degree of abrasion and no doubt they are natural, being the internal section of gourds. These gourds are dried, their outer walls removed and the naturally brown rough inner bleached before they hit the shelves. It need not be said that where a process involves bleach it does detract somewhat from the eco-friendly credentials of the product. Other natural fibred alternatives include coarse fibres such as coir, which is the hairy outer covering of a coconut. Coir is already in use for products such as doormats, but it does have a habit of breaking and shedding. Many other materials, of course, are being investigated including jute, broomcorn and hemp. So really it is just a matter of finding the balance between functionality, ethical production and environmental considerations. It really is an individual choice.

That choice extends to asking whether we even stick with the humble pot scrub? Or maybe we should shove everything into a dishwasher and worry later about energy use, not to mention adding quantities of detergents and salt to the water system. As an aside, it might be interesting to mention those detergent pods sometimes used with dishwashers or even washing machines. The liquid component itself usually contains enzymes, surfactants and even bleaches to get everything grime-free and squeaky clean, but the pillow-like squidgy pod itself is likely to be made of poly vinyl alcohol (PVOH), itself derived from poly vinyl acetate (PVA), which maybe is most encountered as a glue. Both are of course a polymer and highly processed, but in this case the PVOH casing can dissolve in the machine becoming CO_2 and water. So yes, it is another source of CO_2 – tiny, of course – but it does not form microplastics. So again here is a question for the individual: keep things simple with a pot scrub and squirt of washing liquid, or switch to block detergents and a dishwasher? There really is no right answer, both have their pitfalls. Even washing-up liquid contains surfactants and other elements maybe not ideal to enter a water course. One thing is for sure: we can't go back to river sand and reeds.

Wet Wipes

Wet wipes are a strange product. They, like bottled water, have cropped up en masse relatively recently. Invented by a New Yorker named Arthur Julius in the late 1950s, he devised a method for packaging disposable wet napkins. In fact he called his product 'Wet Nap', so reflecting the unique selling point of these tissues. From these early beginnings, a productive meeting between Arthur's son and the legendary 'Colonel' Saunders in 1963 saw these wipes deployed across a certain well known fried chicken fast food chain. Arthur's company, Nice-Pak, went from strength to strength and is still going strong today. The watershed moment perhaps came in 1992 when they entered the UK market targeted at the baby sector.

So, from their original mass deployment contained within little lemon-scented sachets added to a fried chicken takeaway order, the next big marketing push was towards the baby sector, where they were advertised as an effective method of wiping bottoms that had been cocooned in dirty nappies. This was the market where they really found their calling, yet there was nothing miraculous in their function; they didn't eliminate entrenched baby grime untouched by other products. In fact, they exploited an odd niche in the market. Rather than dipping a piece of cotton wool or flannel in water, wet wipes offered an alternative. Pluck out an already damp disposable cloth and use, so saving all of two seconds. Like cotton wool it was disposable but you didn't need the fuss of water. While this may be handy when on the go, most if not all baby-changing facilities do have sinks in close proximity to the changing area.

To keep the cloths moist a mix of chemicals are used. Over the years these chemicals have been found to be very effective at cleaning a whole range of products, from car interiors to entire bathrooms. A study looking into the cocktail of chemicals used discovered an average presence of twelve, including such tongue twisters as phenoxyethanol, disodium cocoamphodiacetate, disodium ethylene diaminetetraacetic acid, propylene glycol and iodopropynyl butylcarbamate.[36] Of course, the chemicals are safe – all have to pass stringent tests – but it does seem anomalous that polycarbonate baby bottles are castigated for containing BPA yet these wipes are freely used on the fragile skin of a baby with little questioning. What about scented wet-wipes? Here we can add another chemical to the ingredient list. If the wet-wipe is fragranced it could very well contain phthalate.

There is a far greater and more pressing problem, however. This 'convenience' product is the plague of sewers up and down the country. The Thames reportedly[37] has great underwater banks or reefs made up largely of wet wipes. This is because wet wipes are repeatedly flushed down the toilet in the mistaken view that they will decompose. The vast majority will not, as the base cloth is made of polypropylene and/or polyester. It will behave similarly to a flushed plastic bag. Maybe we need warning labels across toilet seats reminding people that the toilet is not a portal for throwing away our rubbish. That means no wet wipes, or cotton buds for that matter, allowed.

CHAPTER 12

Toothbrushes

Let's take a closer look at another common but easily overlooked item, the toothbrush. Of course there has always existed a desire to maintain our teeth as best we could, and in earlier times this involved rubbing our teeth with a cloth dabbed in a coarse, abrasive substance such as salt, ash, soot, ground shells or charcoal to scour off unwanted residue. Frayed twigs or even roots could also be employed and either rubbed on the teeth or chewed.

In the 1770s an Englishman by the name of William Addis found himself in prison for inciting riots. Wishing to keep himself looking presentable and not keen on swilling around his mouth with a salty rag, he found having extra time on his hands was conducive to thinking up viable alternatives. He devised a functioning toothbrush made from a carved bone handle drilled at one end to embed hogs bristles which were held with wires. He found this worked very well, so continued to develop the idea until the 1780s when he started mass production. His business went from strength to strength, adapting through the centuries to launch the first brushes with nylon bristles in 1938. In fact, Addis's company is still a familiar name today, trading as Wisdom Toothbrushes.

Modern toothbrushes – and here we're first just looking at the manual variety – are relatively small, in twice daily use (hopefully!) and despite their size can be made of three different plastics. As all three material types are brought together very effectively, it does make reprocessing challenging, which is a pity, as the nylon bristles, PP handle and TPE (thermoplastic elastomer) grip are all materials perfectly capable of being recycled.

Here is a prime example of where a conscientious consumer can be met with a dilemma. This is a product that is problematic in recycling terms, is in widespread use and needs replacing regularly.

What are the alternatives though? There is an increase in those who have chosen to eschew all things associated with animal derivatives. Many of these individuals are also environmentally aware and strive to make the best choices for the planet. But in such a case what is the better choice? Even where we switch to the electric brush, we retain all the aforementioned materials but now add in metal to the construct. Of course we dispose of a much smaller portion when we renew brush heads, and there are strides being made into recycling these, but problems separating polymers do remain.

Using polymer-based toothbrushes enables us to maintain oral hygiene without involving the useful bits of an animal. Using more traditional materials such as those adopted in the past, including bone, hog bristles, bamboo or wood will not at this moment in time give the same degree of oral hygiene as can be found with plastics. Their efficacy may not quite be as we hope for in a toothbrush, and we can encounter similar problems mentioned in many other products using bamboo, etc. – and that is, to make them usable, strong, splinter-free and water repellent, a common solution involves the use of a polymer adhesive.

Here we are presented with an interesting problem. Do we re-examine the materials used, acknowledging that, certainly for electric versions, the solution might not be so clear if we wish to eliminate polymers. Might this product instead call for an interesting design solution? Maybe we can investigate a snap-fit form allowing for ease of deconstruction post-use. Maybe we can re-examine the materials used and ensure we adopt those that can be recycled together. It may be a small product, but it all adds up when we want to keep as much as we can from landfill.

CHAPTER 12

Banknotes

Polymer banknotes were introduced to the UK in 2016 to mixed review. They are made of bi-axially oriented polypropylene (PP). Basically this is a sheet of PP which has been stretched at right angles to itself to make it clearer and stronger than its naturally opaque form. PP has cropped up many times now. It has some very adaptable properties and excels under certain conditions. For instance, it has terrific flexural properties – in other words it can be flexed or bent back and forth many times without breaking or cracking. Although many materials in film form have this trait, it certainly adds to the material's usefulness as a banknote, particularly when that banknote is destined to a lifetime folded into purses and bent into wallets. It will suffer little detrimental effect no matter how many times it changes hands. In fact, this ability to recover from folding remains an annoying trait in these notes which, released from the confines of a wallet, seem to retain their perky attitude and ping up at you any chance they get.

Of course, many interesting and clever additions are added to the base material to keep it secure from counterfeiters and to make it appear right for its application. But fundamentally PP costs the bank a lot less to produce and lasts longer – especially if it accidentally runs through a wash cycle in the pocket of a pair of jeans. Survival in a wash cycle may be a good thing in retrospect. All manner of germs can be passed on through banknotes, and traces of many unsavoury substances have been found on them, so the fact they can now be wiped down may be an advantage. While we are on the subject of ease of cleaning, the products where we probably fixate the most in terms of cleanliness are toys. Hygiene is certainly a major advantage when it comes to toys, but there are many other advantages when our thoughts turn to children and their entertainment.

Toys

An average house containing young children is highly likely to have toys strewn about pretty much every available room at some point in the day. Toys are very important to a child's development and a very interesting use of plastics. Although there is a certain backlash promoting more traditional materials for the manufacture of toys, to a large extent they remain manufactured from polymers.

There are several reasons why, but before considering these, why not head back in time to say one hundred years ago. Then, and in reality up to comparatively recently, toys were very gender-specific. Girls had their dolls and boys had their toy soldiers. The dolls comprised bisque or porcelain heads and bodies for the rich, lacquered composition dolls made of a mix of glue, wood flour and sawdust as an inexpensive alternative for the middling sort, and rag dolls for the poor. Either of the latter may have varying quantities of cloth, sawdust or straw stuffing.

Boys also had a choice in the quality of their toys with cast lead-painted soldiers for the rich, and again a composite alternative as the cheaper option. Wooden toys of course were available to all and often abounded. Supplementing these gender-specific toys, soft toys such as teddy bears could have rabbit fur outers and be stuffed with sawdust, straw or horsehair. Tinplate was common too, especially for wind-up toys.

These toys were by and large expensive. Children of families with higher incomes gained access to the full extent of these delights, often in quantities determined by their wealth. However, children on the whole, whatever their background or price point, were careful with their precious toys, a definite plus point for bygone times. Not so positive was the sheer delicacy of some of these toys. Paint chipped off tinplate and even wooden toys.

Porcelain or bisque dolls could shatter, and anything stuffed with straw or sawdust, or made of animal fur, posed their own problems if they needed a wash or had an unfortunate encounter with a moth or vermin. Lead of course threw up problems of its own, with a turning point coming in 1966, at which time laws banning this metal from many uses curtailed manufacture, in particular of the cast toy soldier.

The advent of plastic was revolutionary for children. Toys could be manufactured cheaply, robustly, in tactile shapes and appealing colours. These toys did tend to be brightly coloured, particularly those toys destined for the littlest fingers. Bright primary colours are eye-catching and advantageous for the child's development. Here the colour is dispensed as pigment throughout the toy and not painted on as would be the case in traditional toys. This lends a certain durability to the lifespan of the toy, not to mention giving a decided safety feature. There is a reassurance knowing that a child can gnaw happily at a toy without paint peeling or flaking and otherwise being ingested. There is another advantage of course: having a pigment dispersed evenly throughout the toy avoids loss of colour when that toy is scuffed or scratched.

Of course, if a toy has seen the pavement, the inside of the dog's basket or sundry other unsavoury sites, there is a certain reassurance in knowing that plastics toys by and large can be hygienically cleaned. These plastic-based toys can take a bashing, be flung about the place and display a durability not seen in many other materials. They can also be easily toted about thanks to their lightweight attributes and are unlikely to present little fingers with sharp edges. All in all, plastics have revolutionised the toy industry. Maybe too much.

Here again we seem to have gone into overdrive. It is hugely beneficial to all having turned our backs on the elitism previously attached to toys. Certainly, well-made and well-designed toys can really help the cognitive and motor functions of developing children, but when that child is offered a toy each time they are taken for a fast food meal or given a chocolate egg, then the joy experienced on receiving a new toy is dulled to the extreme. It cheapens the experience and overloads the child with; well let's face it, junk. It is yet another marketing ploy targeting the youngest in society conditioning them for a lifetime of brand loyalty and consumerism. This is just the generation who will grow, and indeed are growing, to wonder how we managed to accumulate so much plastic waste. It is this waste that is likely to be shunned by the recyclers as the 'hard' plastic many refuse to take. It all of course can be recycled, but it can be tricky for certain machinery because of its size and shape.

Maybe these throwaway freebies is a benefit too far. In fairness some chains are curtailing this marketing trick, but there are many others only too happy to fill the breach. We may have moved away from the free gift tumbling from a box of cereal, but we still find these gifts stuck onto magazines and comics. Mostly useless, mostly designed to entertain for only a short time span, and mostly cheap, badly designed and manufactured, they still fill transport containers and chug across the oceans to pop up on shelves and catch the eye of the next child.

Shoes

There is a very strange juxtaposition in evidence at the moment. On the one hand we are being told that plastics ought to be avoided as they are so bad for the planet, and on the other hand that animal husbandry and all products related to this industry is likewise destroying the planet. Veganism is being pushed by retail outlets with an eye for a growing market and the chance to make sales by tapping into ideological sentiments.

This has resulted in conflicting messages. Some stores now market vegan products in the form of shoes, bags and garments.

Indeed one major retail outlet will happily display signs saying how they are reducing their use of plastics while also pitching their vegan footwear. What does this even mean? Certainly it does mean that they contain no leather, even if it looks as though they do. So what are they made of? For the most part – plastic, of course! Once again a product is being marketed to appeal to a certain sector of society. A pair of shoes advertised as being 'vegan' would generate more sales than the perhaps more apt and truthful label saying they were made of plastic.

Leather footwear has always been seen as a mark of quality. It still is. Leather is a hard-wearing, natural material that has been proven over very many years. Most 'vegan' footwear is manufactured from synthetic materials. This means polyurethane, aka PU, in essence the same basic material that makes scouring sponges, in a different form of course. It is in thermoplastic form (TPU) meaning it can be reprocessed, and has other ingredients added to the mix. Can we follow the argument through and say that, if vegan equates to eco, plastics are environmentally friendly? Let's hold judgement for now and instead look a little more closely at these shoes and consider the impact they have made on society.

It is probably fair to say that plastics have shown themselves to be a useful equaliser of social class. We have already glimpsed this when looking at toys. Just a hundred years ago or so, as we have seen, there was a great disparity between those inhabiting different social classes. Keeping our focus on shoes, no doubt we would be horrified to see a child running around shoeless no matter the weather, yet this was a common enough sight in the early twentieth century. Particularly so amongst the working classes who, although they earned a wage, found it was at too low a pay scale for all a family's needs. It sometimes just wasn't sensible to spend hard-earned cash on something with limited use (children's feet grow), when there were other vital calls on the purse strings.

These children, girls in particular, would be destined for work in service, at which point they may be given their first pair of shoes. It wasn't uncommon for this to occur at around the age of twelve. Not all girls enjoyed encasing their feet in rigid, constrictive, and perhaps rubbing, shoes, but they certainly couldn't remain barefoot in their employer's house.

Then shoes came in leather (some as wooden clogs, and some canvas, but the standard shoe was leather). Leather could be durable and hard-wearing for utilitarian use or soft and supple where destined for the upper classes. Leather has to be tanned, worked and stitched. Leather then, as now, cost money. Shoes once bought were closely cared for and maintained by those who had to watch the pennies.

Skip on a few decades and now we are spoilt for choice. Those, even on tight budgets, can have shoes falling out of their cupboards if they so choose. Why? Because now we have a range of materials at our fingertips that simply didn't exist before. Footwear can be made to be waterproof, hard-wearing, shiny, matte, sparkly or any number of additional adjectives you wish to apply, simply because it does not need to be made solely (excuse the pun) of leather. This means there is a large range of possibilities that can suit almost any budget or indeed style. Leather is of course still available and popular, and tends to sit at the upper end of the price range.

Looking at our options, leather is an organic material that can have its own idiosyncrasies, meaning that not all the hide can be used. Using leather on items such as shoes inevitably means a certain amount of scrap. This all adds to the expense of the finished goods.

Fabric has always been an option in footwear, but here also the diversity of materials has increased. Where once choice was limited, now the field has widened and includes synthetic textiles such as nylon, polypropylene and polyester. Each offers their own particular properties. You will see this in shoes, especially sports shoes, that are advertised as being 'breathable' or can give benefits in traction or waterproofness or even support.

Shoes, again particularly sport shoes, can be made up of several parts and as many disparate materials. For instance, the main body of the shoe itself can be layered up with several materials, or it can appear as a simple leatherette type material. This leatherette as produced for shoes tends to comprise a foamed polyurethane (PU) layered onto a backing of woven polyester. The PU can be formulated to a high level of sophistication and can otherwise appear realistic and very leather-like. It can be textured and highly coloured with all manner of additives and degrees of foaming to give an appearance best suited to the aesthetic preferences of the wearer. Wearability can also be factored in, with durable shoes going through a specific adhesive process where the PU is bonded to the backing cloth in such a manner that it will remain smooth and pliant. Cheaper versions will, on continuous flexing, begin to betray each crease and wrinkle pressed into it on every step taken. Either way these PU shoes do tend to degrade at a faster rate.

For those shoes engineered for performance, the polymer deployed is more than capable of imparting just the desired output. By careful foaming of the material, shoes can be made breathable, allowing a degree of permeability to air and moisture. (For instance, PU foamed to twice its volume can make for a good solid yet tractable shoe, foamed to forty times its volume and it makes for a comfortably squishy scatter cushion.) If you remember a few chapters back to where open and closed celled foam was shown to insulate (closed) or transmit (open), this foam can be formulated and tailored to requirements. So where for instance closed cell foam is required, this can mean polyethylene, PU and EVA, as well as certain rubber types. Here we most likely are talking of the soles, a shoe part working best if it is manufactured from a robust material. In all, PU can make for a very useful material in all aspects of shoe production. According to ISOPA,[38] a European group which covers PU, this material is particularly useful in sole production and is more hard-wearing than leather, rubber, PVC or any other material possibly found in shoe soles.

Hidden Uses of Plastics

Now that we have looked at the areas where polymers are in plain sight, why not now look at its multiple usage in areas not tending to be so obvious. Plastics and rubber can of course turn up in all sorts of unexpected places, so it may be interesting to uncover some of these. It has been pointed out in a previous chapter that containers such as drinks cans and 'tin' cans are lined with a polymer, usually epoxy, to protect the metal from the corrosive effects of their contents. We also mentioned coffee jars, or indeed any jar, that also uses a layer of plastic to provide a seal with the lid. And, of course, we can't forget all those shiny paperboard products that are given a waterproofing layer of plastic, such as food trays or disposable cups. These really are plastics hidden in plain sight, but what if we were to dig a little deeper?

There is a microplastic that we purposely manufacture, throw about with abandon, offer to our children and encourage them to liberally scatter. This microplastic is glitter. Its shiny appearance may lead us to believe that it is just tiny shards of some clipped metal, but in fact glitter isn't too different from those practically impossible to recycle foiled confectionary packages. It is for the most part a plastic layered with a metal foil. Usually this will be the ever-present PET or PVC layered with aluminium, although other combinations can provide a similar glitter effect, as can incorporating pigments or reflective elements into the plastic itself.

The glitter can then be sprinkled onto cards or, far more annoyingly, tipped into a card so that it wafts in an irritating arc across the carpet as soon as this bringer of joy is opened. It can also be daubed onto toys, fused into gift paper and boxes or even mixed with cosmetics to smear across your skin or nails. Glitter is not new; even millennia ago there is evidence that our ancient ancestors used mica or other glistening minerals to make their cave paintings glow and sparkle. There is even evidence that crushed iridescent beetles were used. In more recent times, glass was utilised to give that glitter effect, until in 1934 an American rancher by the name of Henry Ruschmann, dabbling as a machinist cutting photographic film, found a method for using chopped plastic effectively as a glitter. Now there is a move towards biodegradable glitter, but chances are the glitter that predominates the high street is of a plastics origin.

The next item on our agenda is the teabag, now the default choice when making tea, even when that tea is made in a pot. Teabags have been around for longer than may be realised. According to the UK Tea and Infusions Association (yes there is such a thing) they were first devised in America in 1908, by a New York tea merchant called Thomas Sullivan. He sent out samples of his products in little silk pouches, and his customers thought the whole bag was meant to be dunked into the boiling water. Once the entrepreneurial Thomas got word of this novel usage for his samples he devised pouches more suited to infusing tea. At that time this entailed a gauze bag. Gauze eventually ceded to paper.

Nowadays, teabags feel papery and pliable but strong enough not to turn to mush in hot water. The paper used in a teabag is actually very well engineered consisting of several raw sources of natural and synthetic materials. This papery material is, like the wet wipe, a non-woven material. In general these bags are a mix of wood pulp and natural non-wood pulps, from which a form of cellulose can be extracted. These include manila (think brown envelopes), which is itself a form of hemp (as used in hemp ropes). This fibre is relatively expensive, but does help in giving the desired strength to the bag. Added to the mix are the synthetics such as polypropylene and/or polyester. Both can act to strengthen the bag, with the PP content making for a handy and easy means of forming a seal and encasing the tea leaf filling. Of course, the fact neither softens nor melts when boiling water hits the bag is another plus point. This layered-up paper generally contains plastic across the entire bag, despite really needing it only for the edges. It is a more cost-effective way of incorporating the seal.

Of course there are now other options coming onto the market, but as the new so-called bio bags (in very many cases) consist of PLA (polylactic acid) – again a material we have already encountered – this may not offer the panacea desired. PLA is indeed created from a bio source, usually from maize, but the methods used to create this PLA do not necessarily make it easy to subsequently degrade. The best it can hope for is to be industrially composted in conditions not met in a back garden compost bin. The best we can do at home is split the bag and empty out the used leaves, or just use loose tea in the first place.

A last note on teabags, not only can we find these packages in plastic-wrapped boxes, but where they are individually wrapped we have a plethora of other materials to deal with from the individually sealed sachet (sealed almost certainly means via a plastic layer) to the little paper tags which can be found stapled onto a dangly thread linked to the bag. This is overthinking a simple cup of tea and introducing a difficult-to-separate multitude of materials to the job.

Staying in the food and beverage sector, let's have a look at sausages. Of course the early forms of sausage used a natural casing. Traditionally the small intestine was the go-to encasement. This gossamer like covering can still be found, but now is joined by a number of other possibilities. These include collagen, which can be derived from the hides of animals, cellulose of plant origin, textiles such as linen, and wholly synthetic materials such as PP, PE, nylon or a mix. The dry deli type sausages would normally be sheathed in a material that is obviously not meant for ingestion and can easily be peeled away. Bangers in the pan can very much be a mixed offering. It is not hugely uncommon to find natural polymers acting as casings, with alginates derived from seaweed for instance being used, particularly for the likes of vegan sausages.

Staying in the realm of food, chewing gum is an eye-opener when it comes to polymer content, since it pretty much only consists of sugar (real or artificial), a bit of flavouring and rubber. When I say rubber I really do mean rubber. The chewy component keeping those jaws masticating is a common rubber called polyisobutylene (PIB). It's perfectly acceptable, in terms of food standards, to be used in such a way, as it is not a toxic elastomer.

This particular synthetic rubber can commonly be found in inflatable inners because it is particularly good at retaining gases such as in air. PIB is a useful and versatile material and in varied form can turn up in all sorts of applications, from adhesives to sealants to fuel additives, from cling film to shoe soles. Rubber may not be the immediate thought when chomping on a stick of gum but that is the case for the majority of gums out there. They are certainly not likely to come from the sappy tree resins where they originated.

This knowledge may make litterbugs think twice when flicking it onto the street. Just because it can be ingested doesn't mean it will degrade. On a better note, though, it can, and indeed sometimes is, recycled. Anyone spotting those little pink gumdrop globes fastened to lampposts in railway stations or on the street may not realise that the gum dropped in is ground down and moulded into other products. Indeed the gumdrop container itself is made from recycled gum, meaning that the entire thing can be popped through the recycling process without the need for anyone to attempt peeling off the sticky gum.

A related polymer, polyisobutene (there is a very subtly spelling difference indicating the slight change in the material) also crops up in unexpected areas. When a slick of lipstick or lip-gloss is applied, chances are it will contain this material. The reason being relates to its waterproofing capabilities, not to mention its ability to impart a healthy glow. It can keep the skin's moisture in and other moisture out. Face cream and other make-up products can also contain polyisobutene, as can sunscreen. Sunscreen in fact can be pretty chock full of polymers of varying descriptions. Take a look at a label. Many of them contain a plethora of ingredients, most synthetic and many that are listed as polymers, co-polymers and cross polymers. Not all do, however, so if this is a turn off, please do read the label.

It may be irritating to note just how many places polymers have been shoehorned into, but for the most part they are ably employed and do a the job in hand without causing any issues. Products are complex these days. Few are as simple as we like to think and many contain a cocktail of ingredients to give them just the appearance and function desired. Cost too does play its part.

Silicone

Plastics are threaded through our lives even where least expected. The examples shown already ably demonstrate this. Goods are bought where we may have no inkling that plastics are involved. Even items that have been around a long time may appear on the surface to be deficit of plastics, yet still they can lurk. Silicone is one of those polymers that can show up in the oddest of places, many of these also hidden so this might be a good place to look at this exceptionally versatile material.

Silicones are terrific, highly adaptable materials harnessed for a huge number of applications, many not at all immediately obvious. Silicones are a type of elastomer, in other words a rubber-like material. It is a complex material, with some reports saying it is eco-friendly, non-plastic and recyclable, and others saying it isn't. It actually sits somewhere in between. As ascertained earlier in the book, polymers can be a plastic, a straightforward rubber or an elastomer that sits somewhere in between. It basically is a combination of extracts from sand and hydrocarbons, but it is not a plastics alternative, as some would have us believe, as it is a man-made polymer, does not biodegrade, and although a siloxane (silicone-oxygen combination), it can still incorporate hydrocarbons.

Silicone can appear in liquid form where it can be incorporated into hair conditioner imparting a tangle free sheen to the applicant's hair. Alternatively it can be blended with oils where it adds to the lubricating capabilities of that oil. It can be found in pastes and readily found on the shelves of DIY shops, where it appears in caulks for sealing shower bases or sealants for windows for instance. In cosmetics it imparts a silky soft feel to products particularly when they are applied to the skin. In medicine it is found in multiple areas from tubing used in catheters, drains and shunts, seals in syringes, elements in heart valves, some contact lenses and certain implants.

In mouldable form if produces the flexible teats found on baby's bottles, the soft keys depressed on phones or remote controls, in the kitchen as spatulas and pastry brushes. Where heat is encountered it makes for good grippable oven mitts, slotted spoons for cooking and dishing out a fry up, or coated mats and baking trays for sliding biscuits out of the oven. It even makes for good bakeware particularly where fancy shapes are needed, think heart shaped buns or bear shaped cakes. Where cold is encountered it makes those handy flexible moulds for making shaped ice cubes, or just parts on containers destined for the freezer. In particular it makes for a great seal, and so can be found in bead form embedded in the lid of PP snap-fit containers. The list really does go on.

We have now examined a variety of products and in some instances, followed them from their older to newer form. Next we will take some specific examples as case studies and take a scientific, research-based approach in considering whether plastics or a more traditional material choice is better.

13

GO COMPARE – CHOOSING PLASTIC OR AN ALTERNATIVE

We have covered quite a bit of ground in these last few chapters with a little more context given to the myriad of materials we may choose to use. And offering choice really is at the heart of much of the information given. Now we have certain salient facts on polymers and their alternatives at our fingertips, some miscellaneous yet representative examples are well worth closer inspection. Some may surprise when looked at from an environmental perspective, and not exactly conform to common perception.

Christmas trees

Let's start with the Christmas tree – which camp are you in: real tree or artificial tree? There are a whole host of reasons for choosing one above another, but can environmental concerns add to our selection process? As with practically everything else, this isn't as straightforward as it might seem. There are a whole host of factors to take into account, so unless you are digging up a tree from your own garden and then replanting it on Twelfth Night, it might be a surprisingly difficult choice.

Although we may intuitively reach for a real tree if we wish to be environmentally conscious, the real tree option surprisingly is not without its own environmental problems. Much of this has to do with carbon sequestering and release. We all know that trees are great for sequestering carbon, but it is just that, sequestering, storing it away. It doesn't convert into something non-carbon. CO_2 is taken in, broken apart, the oxygen released and the carbon remains. As soon as a tree has been cut down it begins to release the carbon it has stored. The amount and rate at which the carbon is released depends on the fate of the tree post-use, whether it is destined for landfill, compost or incineration. Looking at the abundance of sad withered trees dumped outside many a household post-festivities may give an indication to its fate in many instances. Much does depend on the council policy in a particular area. Some charge to take any green waste, others will offer to take a tree regardless. Some householders will try to squeeze a tree into a wheelie bin; others will bring it to a recycling centre where it can be chipped. The ultimate fate of the tree will make a difference when it comes to the environment.

Perhaps counterintuitively, in terms of carbon footprint, land-filling the tree would be the better option. Both composting and incineration will lead to a more rapid release of carbon. Of course many tree suppliers will replant for each tree felled, but this means a sapling will take the place of a mature tree and so will not have equal sequestering powers until it too reaches the level of growth needed for it to be felled. This of course will take a number of years. Tree disposal is only looking at one aspect in the pantheon of environmental consequences. By the time growing, felling and transport is taken into account, there are additional environmental considerations to be factored in.

Artificial trees are so much easier to identify as having environmental issues. They are made up of several parts and materials, each causing their own problem. It's all a far cry from the original artificial tree made from dyed goose feathers, wires and dowels (the feathers were dyed green, split and wound around lengths of wires which were then pushed into an upright dowel). The first trees to resemble the artificial type we use today were manufactured by a UK company making toilet brushes! They used a similar process to that adopted for the toilet brushes, taking the same bristles, dying them green and twisting them in wire. These could then be arranged onto a central post.

Today there are many more steps to the process, with steel wire and sheeting used as well as PVC, polypropylene and epoxy all combining to make for very realistic alternatives. These need to be formed, stamped, pressed, entwined, rolled, cut and fastened. It all takes raw material and energy. Once the tree has been produced it needs to be packed, usually in cardboard, sealed, usually with polypropylene tape, and shipped. As most manufacture is now in China, this shipping is undeniably also problematic environmentally.

So far it does seem that a real tree is far better on balance. But surprisingly that might not be the best option if one takes into account an additional crucial factor. Artificial trees tend to make an annual trip to and from the loft and get reused many times. The American Christmas Tree Association (yes, there is such a thing!) have looked into this. They say on average artificial trees are used for ten years. Not unreasonable, considering the time span they are in actual use each year. This very much puts the artificial tree back in the frame as being the environmentally better option. A study by this Association suggested that using a tree for 4.7 years would have a lower environmental impact than buying 4.7 real trees over that time span.

There are other reports that put this reuse figure closer to the ten-year mark before that tipping point comes into play, but it really does depend on so many input variables, not least how locally the real tree grew, the fertilizer and pesticide used, how it was harvested and what happened to it post-use. Likewise, many variables need to be accounted for with the artificial tree, including raw materials processed, emissions generated during processing and transportation. A tree is only used for a relatively small portion of the year, so even one purchased from a moderately priced retailer ought to last for many years.

Shopping bags

Plastic bags have come in for a hammering of late. Banned as freebies, they now must be paid for if we wish to avail ourselves of their services. Plastic bag taxes have spread across many countries, with a levy higher than their integral worth pushing us to rethink our habits. It has worked. For the most part we remember our bags when visiting the supermarket and, if short, can easily grab a reusable sturdy plastic bag. It has made a massive difference. The number of flimsy plastic bags in circulation has dropped dramatically. They were light and easily caught by a gust of wind so would dance along pavements, tangle high in tree branches (in Ireland these were known as witches' knickers) or flap across fields. They were a nuisance and few mourned their diminished presence.

But we do need alternatives to carry home our shopping, so what should we choose? Corralled into the same ban were those stronger heavier duty bags used by general retailers. Buy a pair of trousers, and it would be popped folded into a logoed bag; same with electrical goods or indeed most other high street purchases. While we may be getting better at remembering our bags for a supermarket dash, we may not be so forearmed entering the local shoe shop for instance. So unless we bring a suitable bag with us, we need to pay up. While for big-ticket items this might not even register, for some it does grate.

Certain retailers offer free logoed paper bags, so circumventing the awkwardness of having to declare a bag charge. I think it might be no surprise when I say that environmentally this is not the best option we could be given. For practical purposes also, navigating a busy high street while dangling paper bags from our arms does not always guarantee that those purchases make it across the threshold still in those bags – particularly if a shower of rain should chance.

Where does this leave us then? What is our best bet with these bags? Ought we stick with some form of plastic, go cotton or chance paper? Luckily a couple of governmental reports can help in our understanding of the alternatives out there and the impact they have environmentally. One commissioned by the Danish Environmental Protection Agency in 2018 in particular has really delved deep into just this subject and has uncovered some very interesting information, while another one commissioned UK's Environmental Agency in 2011 is also useful.

Both studies looked at vest bags, which are those thin, standard (pre-ban) supermarket end-of-till bags. Of the low-density polyethylene (LDPE) bags available, four types were subjected to the Danish study, including one with average characteristics, which was taken as the base or reference bag. The other LDPE bags studied were those with a soft handle, one with a rigid handle and a recycled version; next in the study were two types of polypropylene (PP) bag, non-woven and woven, then both a virgin and recycled polyethylene terephthalate (PET) bag, then a starch-composed biopolymer (usually polylactic acid aka PLA). Moving to natural materials, next included were two types of paper bags, unbleached and bleached, and then two types of cotton, organic and conventional, and finally a composite bag comprising jute, PP and cotton.

The study wished to identify the carrier bag type that would give multiple use coupled with the best environmental outcome as compared to the simple vest-type plastic bag which is considered single use. So with an eye to the environment, consideration was given to production, distribution, use and disposal.

Certain parameters were implemented, so for instance it was assumed that the carrier bag was made in Europe and so energy consumption specifically reflected this point. The cotton of course comes from overseas, and so energy consumption calculations were based on a global average from China, India, Latin America, and Turkey. Transportation data was assessed with various scenarios spotlighted. End-of-life considerations also appeared in the report and included waste management systems.

The results surprisingly point to the LDPE bag being best (based on the assumption that it would be used once for shopping and once as a bin liner), and this could be improved upon further if it were to be reused even one more time. Each LDPE bag (recycled or with a strengthened handle) fared more or less the same, although the recycled bag really would be best used at least once more (extra processing is needed to recover and reform this plastic). Moving onto PP bags (those non-woven bags for life), they are recommended to be reused at least fifty two times when everything is considered (and to equate it in terms of sustainability to the simple LDPE bag). The stronger-woven PP bags ought to be reused forty five times. Maybe somewhat surprisingly, biopolymer bags were calculated as needing at least forty two uses to equate to the single-use bag. Not far behind were the paper options with both forms requiring forty three reuses. The big shock came with cotton. Taking all factors into consideration, organic cotton bags necessitates reusing a whopping 20,000 times. Considering how we are always steered to believe that cotton trumps plastic, this outcome is remarkable. The remarkably high reuse number for cotton is attributed to its impact on ozone depletion. Omitting this data reduces its reuse values enormously (149 times for organic, 52 for conventional cotton), but remember, the same stipulations were imposed on the other bags, including plastic.

All this really does show how our instinctive selection of a natural and long-established material really might not be the solution we suppose. Certainly it would be an unlikely scenario to find unsightly cotton bags dangling from trees, and if they did end up there, they would probably cause less harm and would eventually break down when exposed to the elements. So this really does take us back to the original argument that plastic is not necessarily bad environmentally but the shoddy way in which It is disposed is seriously problematic and does a great deal to undermine its usefulness.

Having given food for thought with the Danish study, a brief review of the UK study is worthwhile. Here we start with one tweak and that is to the baseline vest type shopping bag. Either LDPE or HDPE can be used, but the UK study went with HDPE. This study also looked at a lightweight HDPE bag with an incorporated additive designed to allow the plastic to break down into small parts. They also considered a biodegradable bag, a paper bag, an LDPE 'bag for life', a non-woven PP bag (the type usually sporting cloth handles sown onto the sides) and a cotton bag. This study also took account of a great many variables, not least being the source countries for each constituent part of the bag, how it was manufactured and subsequently transported. It even took account of interim countries where elements of manufacture took place and the primary energy source available in each of these countries. In all they did what they could to make accurate calculations for the manufacturing processes. Recycling was accounted for too, as was the ultimate destination of disposed bags (landfill or shipped abroad).

The report understood that each bag was produced with differing intentions. Basically some were designed to last longer than others. Those designed for greater robustness were likely to produce greater environmental impacts, so for the sake of a fair comparison, calculations were generated from the number of bags required to carry one month's shopping. A calculation was then made as to the number of times each different type of bag would have to be used to reduce its global warming potential to below that of the standard HDPE bag. Certain other caveats were determined including environmental impacts: such as resource depletion, acidification, eutrophication, human toxicity, water eco-toxicity and smog formation to name but a few. In conclusion the highlights of the report stated that no matter what the bag type, the dominant environmental impact centres on resource use and manufacture, both of which have a greater effect than transport or disposal. Again, no matter what type of bag is used, the key to reducing its environmental impact is to reuse it as many times as possible, even if that use is as a replacement bin liner. In fact, maybe surprisingly, it was found that reusing those lighter plastic bags as bin liners gave greater benefits than recycling those same bags.

It was found that in order to equate to the standard single-use vest-type plastic bag the paper equivalent should be reused three times, the thicker LDPE four times, the cloth-handled non-woven PP eleven times and cotton bags 131 times (each were measured in terms of lowering their global warming potential).

Another surprising find showed that recycling or composting generally produce only a small reduction in global warming potential. Much is made in sustainability circles of biodegrading and composting products no longer required. The goal seemingly being, in this instance, to be able to add a spent bag to a system (such as a compost bin) and the resultant biomass used, presumably in the soil. There is little or no mention of the suitability of the resultant compost to the soil. Not all compost is equal, not all soils are equal.

Say we have a biodegradable bag that can be home composted. How it breaks down and what it breaks down into will depend on the starting product and what materials, bio or otherwise, were used to make that bag. From our viewpoint we may see a brown loam tumble from the compost bin and dig it into our garden enthusiastically. But what about taking another viewpoint? This is where eutrophication comes in.

To quote the Environmental Agencies report, 'Eutrophication is caused by the addition of nutrients to a soil or water system which leads to an increase in biomass, damaging other lifeforms. Nitrogen and phosphorus are the two nutrients most implicated in eutrophication.' So in other words the wrong sort of compost can harm the biodiversity of our soil and can impede the growth of native plants. Where we are home-composting this may not amount to much by way of harm, but industrial compost, especially if it is being increasingly fed with biodegradable plastics, has the potential to be particularly problematic. We have a very fine ecological balance in play and we really do need to be very careful, as we may inadvertently be doing more harm than good.

The solution is to know the right questions to ask. If we arm ourselves with the enough knowledge of the potential pitfalls in a system we can then try to mitigate potential damage. I would say: question everything. This is a tough quandary we find ourselves in and there are no 100% right answers. An instinctive choice might have bent us towards a natural material for our bags, but as we see, none are particularly good options, and even bioplastic does not fare well in comparative studies.

Nappies

There is no doubt that when it comes to problematic single-use plastics, disposable nappies must come pretty high on the list. Their very function precludes attempts to recycle. According to the Waste and Resources Action Programme (WRAP), the UK disposes of around three billion disposable nappies each year. They go on to estimate that each child will on average go through 4,000 to 6,000 nappies until they are potty-trained. These figures are enormous, but it will take a lot to persuade parents to consider the alternatives. This is one area where single-use, utterly unrecyclable plastics seem not to be an issue. Maybe it should be?

Modern-day disposable nappies typically comprise of a layered structure, where the outer layer provides the waterproofing and fastenings, while the core contains absorbent materials topped with a protective inner layer. The absorbent materials are now quite sophisticated in design. They comprise a pulp of bleached cellulose fibre (giving that cotton wool type feel) and a water-absorbent polymer called sodium polyacrylate (SAP). It can in fact absorb huge amounts of water many times own weight. This is why a nappy can go on a toddler and fit nice and snug yet moments later can be swinging about its knees looking as if the child had stuffed a pillow down its front.

While the core absorbs, the inner layer is protective, keeping the baby's skin from the SAP and wicking away moisture. This layer is a 'non-woven' polymer, usually polypropylene just like a wet-wipe. The outer waterproof layer is also a plastic, this time usually a polyethylene (LDPE like the thin supermarket bag) incorporated with integral rubber waist elastics and Velcro-like fastenings. All components are glued together with polymer-based adhesives. Needless to say, these nappies in turn are packaged in PE bags or cardboard boxes.

As mentioned above, the mountain of disposable nappies generated each day just in the UK is an issue. Of course it is. Environmentally though, how much of a problem is this issue and just what practical and environmentally helpful alternatives are out there? We have learnt already that cotton comes at a high environmental cost and in general plastics do not generate such a heavy impact in comparison. It is really a dilemma to know what to do for the best. Wishing to have clear comparable facts available to them on this issue, the Department for Environment, Food and Rural Affairs (Defra), the Environment Agency and WRAP commissioned a report[39] to look into just this matter. The investigation was deep and far-ranging and the results a close call really, with several provisos and advice to parents as to the best steps they could take to make their environmental impact worthwhile.

Let's start with establishing what we have as reusable alternatives. A surprising omission from the study is the basic and simplest old-fashioned cotton terrycloth, so we'll come back to that option later. Instead this study focused on reusable cotton nappies, which for the most part are fitted and integrally fastened. These include shaped, fitted nappies that are pretty much self-contained and fastened via Velcro or poppers; shaped nappies that require a separate waterproof cover – again no folding is needed and fastenings are made via Velcro or poppers; and finally prefolds, which consist of panelled rectangles that are folded to shape and covered with a separate waterproof pants or a wrap containing fasteners. In each case the cloth making up the main body of the nappy is cotton derived. The waterproof overwrap can come from a number of sources including PVC, polyurethane, EVA or even hemp, cotton or wool.

The study delved deep in its analysis. Disposable nappies were considered from their formation. Account was taken of each component within the nappy, from the production of timber and its subsequent pulping and bleaching, to the production of the polymer feedstock and its use in the creation of various components, to the chemical production of component parts of the absorbent beads. In addition, the packaging production for bulk nappies was accounted for and fed into the analysis. Electricity consumption during manufacture was added to the mix, as was the route to the retailer and disposal post-use, whether the nappy was subsequently incinerated or sent to landfill.

Likewise, the reusable nappy was subject to an in-depth analysis. This included looking at cotton cultivation and harvesting, including its fertiliser and pesticide requirements. The cotton's processing from spinning to weaving was then fed into the analysis, as were the additional material components needed, in this case mostly for the overwraps or waterproof pants, but also the liners. The route to the retailer was considered, as well as the consumables needed during use, including detergent manufacture, toilet flushings, water treatment and supply and subsequent sewage treatment.

This study really left no stone unturned. For the reusable nappies, a number of scenarios were assessed. After all, there are individual preferences to consider. The study embraced many of these and so generated results depending on whether the nappies were line- or tumble-dried, the temperature and water usage during the wash, whether the nappies were pre-soaked in a bucket or shoved straight into the machine, and even how full that machine was before setting on a wash cycle. Added to the mix was the UK electricity generation source for the year studied (e.g., if it was coal, nuclear, wind, etc.) and the energy efficiency of the machine used. The disposable nappies were measured for an average across various manufacturers. Consideration was also given to the weight of the nappy post-use. It might be interesting to learn that in those first two and a half years taken by the report to be the nappy-wearing timespan of a child, they are expected to produce 727kg of excrement. They even go so far as to designate figures to the disparate forms this excrement takes. In fact, the report says the figure was determined following a delve into a sample of bins. That must have been an interesting job description. However, this diligence means that this study is so comprehensive in its range that the weight of the disposable nappies can also be accounted for when assessing the energy requirements to transport to landfill. While in landfill the handling at this location was also accounted for, as was any gas generation while there.

The results were close. A lot depended on the variables used. There are a lot of facts and figures offered for consideration, but taking the study's concluding comparative measure of CO_2 equivalents, we can bolt some numbers onto the results. The disposable nappies were tagged with a figure a little under 550kg in terms of global warming impact from CO_2 equivalents (across the two and a half years measured). Taking an average in terms of washer-drier use (used on some but not all occasions) this figure was 570kg for the reusable nappies. However, this could drop to 479kg by line drying, and improved even further if washing and drying efficiently and reusing on a second child. In this scenario, the figures dropped to 370kg. Bottom line, excusing the pun, if a sensible approach is taken to laundering, then cloth nappies fared best environmentally. If low-rated washing machines were used at their highest temperature (90°C), along with tumble drier usage (recording a whopping 990kg), then disposable wins hands down. It really does illuminate how data can be presented to give a very different outcome depending on the angle taken.

Now that we have delved deep into the scientific study I'll ask indulgence for a little anecdotal input. As mentioned earlier, I am surprised that the old fashioned terrycloth was omitted from this study. It is easy to fold, with no great origami mastery involved. They can be folded with slight continual adjustments made as the child grows, and so the nappies used on a new-born are still good for a two-year-old. That can't be said for all the other cloth alternatives. I have road-tested some of these on my own child and by far preferred the terrycloth. I won't pretend that the initial attempts at accurate folding were a great success. In fact, in the first few days the baby from the waist down gave the appearance of being all nappy with a couple of feet poking out the end. The fear of jabbing with the pins and using overly large overpants did little to help and more than one nappy fell off.

It is amazing however how quickly a situation can be learnt from and within days everything began working like clockwork. The first breakthrough came with the realisation that the packet fresh terry nappies were far too fluffy in their newness. A couple of washes would have tamed them but instead an age-old solution was deployed – use hand-me-downs. Maybe unusually the hand-me-downs were actually my own old nappies, saved and stored by my mother for that just-in-case scenario. That scenario came a few decades later than anticipated when they were last carefully cleaned and folded away. However they were perfect. These were the nappies that had been though many washes and had been mangled to the point where the cotton fibres had been subdued to a fine soft veneer. They fitted snuggly and were perfect for a new-born.

As the baby grew the fluffier nappies could be used and all that was needed in addition were the overpants. These did need sizing but were thin and basic. The whole process was remarkably easy. Fold a nappy, place a liner on top, fasten onto baby and pull up the waterproof pants. On most occasions the soiled nappy was only wet, so the liner could be binned and the nappy dropped into a lidded bucket of water with a little sanitiser added, so no smell. Lumpy nappies needed a shake over the toilet and the process from there was the same: bin the liner, drop the nappy into the bucket. In all the only items seeing the landfill was the sheet liner. Nothing needed overwrapping and nothing smelled, honestly. The nappies were then washed every second day (on a full load) and dried on the line or clothes horse, all depending on what the weather was doing. Being just a square of terrycloth they tended to dry pretty quickly.

At one stage I had been given a set of prefolds. These were panelled, with the central panel being the thickest. They were of a woven cotton and came with a couple of Velcro fastened wraps. They were certainly nice and easy and didn't need pinning, but the child soon outgrew the wraps and the thick woven prefold took an age to dry. Certainly they have their uses, but in terms of sustainability I really would advocate the simple cloth square. I bought one pack right at the start, and these were supplemented with my own old nappies and they lasted the duration. They might have been an odd choice for a post-millennial baby, but they worked perfectly, no pins ever jabbed skin, and the nappy fitted absolutely perfectly each time and never developing an unseemly sagging about the knees. And now those nappies are carefully washed, folded and put away in readiness for the next generation.

Straws

Before we end this section, one additional product is worth a little scrutiny, and that is the straw. A ban on plastic straws in the UK came in April 2020, but due to the pandemic, it had not been widely noticed. The public, however, found they still wanted straws. Just because they couldn't be made from plastic didn't make one jot of difference, a straw was expected, particularly when purchasing a drink from a fast food outlet. Paper stepped in to fill the breach, but soon it was discovered why plastic acquired the mantle of straw-making in the first place. It wasn't long before complaints were voiced. Paper soon went to mush, collapsed in the mouth, stuck to your lips, left an odd taste and bent in the drink. Some resorted to using several straws where before one had sufficed.

It was a strange item to ban in my opinion. Certainly they are single-use, and that is always going to be an issue, but then so are paper straws. In fact paper straws more often than not can't even last for a single use. I have seen paper straws collapse, especially in the mouths of children, who then replace their squashed straw with a fresh one. Indeed I have seen waiters pre-empt this collapse and offer a handful of straws per drink. How is this in any way better, and just what was wrong with plastic straws?

Plastic straws are eminently recyclable, being made from polypropylene (PP), and in my opinion there is no reason why these straws can't be put through a scrap process system and recovered. Food residue is always an issue when looking to recycle plastic, but just how much residue is on them? They convey liquid, so ought not be any more of a problem than, say, a plastic milk bottle to recycle, residue or not. They aren't blended or layered, and they don't even come into contact with anything that may prove hazardous further down the recycling process. They aren't in tricky film form that binds machinery, nor do great snaking lengths of straw entwine machinery. They are short and robust. Why were they not recycled and why were they not the subject of a campaign to promote recycling?

The reason most commonly cited for not recycling straws relates to them falling through grids or off conveyor belts at recycling plants – so develop better methods of collection. Recycled PP has a commercial value, to see straws washed up on beaches around the world is a disgrace and utterly unnecessary. If there is such a major flaw in our global disposal/recycling systems then it must be addressed and not brushed under the carpet by banning materials that will highlight this flaw. We really must stop ignoring the fundamental problems existing in the refuse sector. The question ought not be, how can we stop this material existing, but how do we stop this waste?

Also, what makes a soggy paper straw better? Soggy paper straws are difficult to recycle to say the least, but now another problem looms. Ever entrepreneurial, manufacturers are circumventing the soggy issues: if we couldn't go back to plastic, we needed stronger better paper straws, and now they have been developed, job done. But is it? What is making these straws stronger? Glue, I was told when asking a manufacturer. But what is this glue? A polymer, of course, except now we are making paper rigid by a coating of non-recyclable plastics-based glue that is so impregnated into the paper that the paper can't be recycled either. All we have achieved is swap a straightforward, perfectly recyclable material with a mixed material, still containing polymer, but now not recyclable. How is this better? While there are paper straws that contain soluble glue, this of course means there is a payoff in terms of sogginess and speed at which the drink must be consumed. A good strong straw unfortunately needs a strong and waterproof coating to bind the paper.

We as consumers have a choice of course: not use a straw, use a paper straw, use a strong polymer-infused paper straw, use a steel straw or use a natural straw such as bamboo. Of these, steel straws are increasingly found in the shops. Stiff and unforgiving, they'll do the job. They are reusable certainly, but will through use begin to harbour bacteria and other unwanted sediment. Therefore many of these straws are sold with a handy little bottle brush that can be inserted into the straw to sluice out anything untoward. These little brushes are made from nylon bristles twisted into a steel wire and sometime terminating in a plastic grip. Now we are back into those tricky-to-recycle, combined disparate materials – not to mention the impact steel production has on the environment – all because we didn't want a plastic straw. Other straw alternatives are coming onto the market, but none are without environmental impact. Where a natural product seems unfeasibly uniform, smooth and blemish-free it may be worth probing further to its manufacture. Is it just a cleaned up version of the raw material or has anything been added to give it the desired function? Remember, if it seems too good to be true, then it probably is. So, just like the bamboo cup example given several chapters ago, some straws are becoming more binder than natural material, and on numerous occasions this binder is a plastic. Where plain unadulterated bamboo is used for instance, then these too have issues. Being natural they can break down, fissure, harbour unwanted bacteria, take on strong odours. Then consider where they originate, how are they treated to appear as they do on the shelves? How were they grown? Has demand meant a move to monoculture in their geographical area of origin?

While it is out of the question that we sit back and do nothing, a little forward thinking would not go amiss in these circumstances. If a product is banned a void is created. How will that be filled and what will be the outcomes? This really does apply across the board. Next time headlines foretell of another ban consider, will this have the desired intentions or will it only divert attention from the true problems? Lobbying may be a solution. Request that action is taken to ensure all waste is dealt with correctly and effectively.

14

CLEANING UP OUR ACT

While plastics have done much to improve our way of life, there is no denying that it has also caused incredible challenges not least to the environment. These must be addressed if we wish to continue our reliance on plastics, in whatever form they take and from whatever source they emanate. There are two areas in particular where a great deal of concern is voiced. This relates to the formation of microplastics and the presence of plastics in the ocean. Both must be addressed if we are to consider polymers in their entirety and particularly if we are to be confident of their future presence in our lives.

Sorting the micro from the macro

Microplastics are of great concern to many, yet there is no clarity in just what we mean by the term or how we define it. There is some confusion when classifying this designation, with official sizes spanning anything from several millimetres in diameter down to truly microscopic levels. This is a massive disparity, so it would be helpful to pinpoint areas of greatest concern. Pieces measured in millimetres will impact far differently from those that can only be seen through the lens of a microscope.

So what is a microplastic? Basically, it is a little piece of plastic that is usually reduced in size through external action such as attrition (this is not always the case, so we'll come back to this point in a moment). Exactly how little is open to debate. Some studies count pieces as large as 10mm, while others refer to particles in the region of 1mm or less. Increasingly microplastic classification refers to pieces less than 5mm in length/diameter, so this can be taken as the norm. (Confusingly, this is not really a micro size as it is scientifically understood, this refers to a measurement of one millionth of a metre. Bringing that down a little more, that's a thousandth of a millimetre or, in other words, 0.001mm.)

Truly microscopic particles of plastic are now also being recognised as a potential cause of concern. These are the nano-plastics that are measured in the nano range or in other words to one billionth of a metre. Research on particles of this minute size is relatively new and is ongoing, as indeed it is in the other size ranges such as the yet smaller pico size. This is now reaching for atomic level, with a picometer being one trillionth of a metre.

Whatever end of the scale, the issues posed by each variety differs fundamentally, so it is important to be able to distinguish between them. By understanding the very different ways particles behave, we can begin to tackle the issues they present to the environment and to us.

Midi-plastics

Midi-plastics are those larger pieces of plastics that break up and find their way into habitats they have no right to be in. Sometimes they consist of broken pieces of plastics that are too fiddly to bother about on a litter pick, but aren't quite the 5mm or less that the majority of commentators specify as being a microplastic. Nevertheless they do need acknowledgement. They can be more than just chips of broken plastic. As mentioned a few chapters back, some so-called biodegradable plastics are actually designed to break up into small pieces. As long as the plastic breaks up to leave 10% or less of the original, then certain standards will allow that the whole has indeed degraded. The 10% remaining of course becomes a microplastic or sometimes even the larger midi-plastic. There is a hope that those biodegradable plastics will further degrade, and some will, but there is a chance that they won't if the conditions aren't right or if they are the wrong type of degradable plastic.

Certain degradable plastics are designed to decompose by a process known as oxo-degradation. Oxo-degradable plastics are conventional plastics, like the HDPE or LDPE-made carrier bags, but with certain additives present. These additives are designed to react with oxygen or UV light to break apart. As it is the additive that decomposes and not the plastic, they will by default leave little islands of plastic that may attrition over time or break apart via UV attack, but they will not disappear into a form of biomass. These little flakes of plastic will remain fluttering about despite the best intentions of consumers thinking they are doing the right thing by attempting to compost. Their problematic nature has been acknowledged by the EU, which has put this form of degradable plastic on their ban list.

A more immediate problem and one proving to be the bane of many a creature are those larger pieces of debris that can choke, strangle and entangle. It may cause *us* no immediate harm but is a plague to the creatures caught up in it, many of whom are sea dwellers. A lot originates from netting and such jetsam as is flung from fishing vessels and other boats. This undoubtedly is in need of serious attention and will be considered in a following section (looking at oceans). First though we'll continue our examination of microplastics.

Microplastics

We have already established that microplastics are those tiny pieces of plastics typically of less than five millimetres. How they can arise may be surprising and best explained in their categorisation as primary or secondary microplastics. Primary microplastics are those which are directly formed, in other words they are deliberately made. These microplastics are surprisingly common, made for example to provide abrasion or exfoliation when added to facial scrubs or other invigorating body-care products. These beads were added to cosmetics, soaps, lotions or pastes to clean, buff or exfoliate anything from the soles of our feet to the enamel on our teeth. Of less than 0.1mm in size and likely to be PE, PP or PS, the UK government imposed a ban on these in 2018. They revealed their findings, showing just one shower using these products was thought to send 100,000 microbeads down the drain. These beads can also be used on an industrial scale as blasters to surface-clean machines, equipment or even boat hulls from rust or paint. Here the beads are more likely to be the made of more robust material such as acrylic, melamine or polyester.

Other primary sources abound in products such as pesticides, where small capsules are designed to control the release of active chemicals when applied to crops. Agriculture can also make use of PE40 as an anticaking or controlled release ingredient in fertilisers (The EU Fertilizing Products Regulation will ban this usage in 2026). Primary microplastics can likewise appear in the medical sector and of course in many a craft box as, for instance, glitter, beads or colourful stickers and embellishments.

Further sources include those expanded polystyrene (EPS) beads appearing in beanbags or, in greater mass, within cavity wall insulation. Injected into walls many escape and dance away on the lightest of breeze. They can escape through gaps in mortar or openings in the eaves. EPS beads are terrific insulators, but has any thought been given to these beads post-use? What if the house is undergoing demolition, or has the misfortune to have its structure compromised through storm action, accident or worse? EPS is light, very light. Once released, beads can easily be swept away. Has any thought been given to containment in such instances?

Other purposely made yet problematic microplastics include artificial turf infill such as that used in artificial football pitches. This in-fill includes pelletised car tyres, which in itself is a commendable approach to reusing these difficult-to-recycle products. Old tyres became particularly difficult to deal with since a ban on landfilling scrap tyres came into force. However, purposely reducing them to microplastic dimensions and applying them to an area where they can be washed away or otherwise enter the environment, maybe through clothing, has raised some eyebrows (anyone with a child kicking a football around such surfaces will appreciate just how much can end up in clothes and shoes post-match). Again it is a case of trying to strike the right balance. The European Chemical Agency (ECHA), an EU body, states that granular infill such as that found on artificial football pitches or playgrounds constitutes a substantial source of microplastics entering the environment, with estimated annual releases of up to 16,000 tonnes. There is also a question mark over the toxicity of these microplastics when considering the chemical composition of certain originating tyres. They can potentially leach chemicals or volatile organic hydrocarbons. To safeguard against potential problems, the EU has announced new rulings to curb certain chemicals entering the environment in unacceptable levels; these came into force from August 2022.

Secondary microplastics originate from the degradation of larger plastic objects. These are those bits that get knocked off a multitude of items from bottles and buckets to pipes and potties. They wear down, usually through mechanical action or UV attack. There is now a greater awareness of how our clothing can contribute to the formation of microplastics. So much of what we wear is synthetic, so it is maybe no surprise when each time our clothes go through a wash cycle we inadvertently cause tiny pieces of (synthetic) fibre to loosen from these garments and enter the wastewater. Perhaps a washing machine manufacturer might design in a trap or filter that could gather and contain these fibres.

Nanoplastics and smaller

The field of nanoplastics is a relatively recently observed state when considering plastics. As research continues, it becomes increasingly certain that if a plastic can degrade to micro size then why would it stop there? It is far more likely to continue reducing in size until truly microscopic. The question is whether at these sizes plastics are a risk. Indeed, at what point do they cease being plastic and become base chemicals? Plastics are held together through molecular bonds. These bonds can come under attack, as mentioned earlier, through UV exposure, oxidation and even the action of light and water. Certain plastics contain additives, not all of them pleasant, so at what point might these be released and cause harm? The very size of these particles makes research tricky, but it is an area beginning to be investigated.

Microplastics are found in growing quantities in the environment, including, of course, our oceans. According to UN calculations there are as many as 51 trillion microplastic particles in the seas. They crop up in all manner of food and drink, and, maybe unsurprisingly, plastic particles have also appeared in human excrement. Their effects on us as yet remain an unknown entity. Certain additives may contain harmful substances, but can they be absorbed if the host plastic is ingested? What biological effects can we expect from microplastics?

Biological effects

A lot of media attention on the subject centres on marine creatures ingesting microplastics, which we in turn purportedly ingest.

Sweeping and emotive statements are made. Where plastics that are greater than a true micro size are ingested, these particles can't pass into the blood stream. They are in the gut, where they may stay or pass through. Likewise if we inadvertently swallow a piece of grit, it will pass through us. Any plastics found in the guts of fish will be removed before we eat them. We don't tend to eat fish guts; this is a waste product. However this isn't our primary concern: we of course are mindful of any foreign bodies entering other living beings. We need to ask, do these particles give off harmful chemicals? It would be especially alarming if the host were to be chemically affected and subsequently has any adverse effects up the food chain. It is not so simple to give a definitive answer. Plastics for the greater part are inert. Categorical proof of chemicals emitting from microplastics just doesn't seem to be out there, nor can we say for certain that any chemicals can pass into organs. If plastics are safe in larger sizes, what alters to make them unsafe in smaller sizes? Maybe their greater surface area could result in a greater potential to leach? We have yet to establish if this is the case. Much more research is needed to establish with any degree of confidence if microplastics are harmful.

The European Chemical Agency (ECHA) has instigated a discussion in this arena expressing concerns on the potential effects of microplastics both on the environment and public health. As part of their current strategy, they established an evidence-gathering timetable, incorporating a public consultation in 2022 followed by a scrutiny of their findings in the EU Parliament. From here possible restrictions will be adopted. Of course the UK is no longer within the EU, but any findings are likely at the very least to be considered in the UK also. Of the research already conducted (at this point across several years) scant concrete evidence has emerged. A review in 2016 for instance considered the presence of microplastics and nanoplastics in food and seafood. Little by way of data accrued from microplastics study was found, and no data at all was available for nanoplastics (in some cases findings declared the presence of these particles but not their potential effect on health). This review, undertaken by another EU group, the European Food Safety Authority (EFSA), recommended more research.

Research did continue and data was amassed; then in 2021 the group held a scientific colloquium in to discuss and share findings to date. Again the focus centred on the effects of microplastics particularly when ingested in food and drink. Several studies on the subject were presented. One in particular addressed the potential health risks emanating from chemicals within microplastics. They identified those chemicals present across a number of commonly encountered plastics (such as PE, PP and PET among others) and examined whether these chemicals could be released into the intestine and if so was intestinal absorption likely. Of the organic chemicals identified, the study showed emissions were very low and few able to translocate across the intestinal barrier. Of the other elements found, there was scant evidence of release and practically no movement across the intestine. Anything present was at levels far below tolerable daily limits.

An additional study then considered the presence of nanoplastics in food and thereby the human digestive system. It was ascertained that plastics of less than 150 μm (or 0.15mm) could be absorbed, but that only plastics of less than 1.5 μm (0.0015mm) could penetrate through human organs. However, any nanoplastics found in the intestinal wall were of very low concentration, and no plastics were detected in the organs examined. Again it was advised that studies do need to be continued. While it may be foolhardy at this relatively early stage to say no biological harm emerges from these plastics, there is no getting away from the fact that the vast majority of scientific studies to date show they do not present a major concern.

So why do we immediately assume the worst when it comes to micro- and nanoplastics, why assume they are so detrimental to our health when no real evidence to back this statement seems to exist? Maybe it just makes for a good story. While the narrative around plastics depicts them as the enemy causing no end of problems to us and the environment, can we say this is a fair or balanced assessment? An interesting joint study by the Universities of Plymouth and Vienna found that while just 24% of scientific studies suggest ill effects arising from micro- and nanoplastics, a whopping 93% of media articles on the subject imply they carry a risk. This probably explains why the public express such concerns which, so far the science is just not realising.

The media remain keen on printing an eye-catching headline. The potential of plastics to migrate through the body via microplastic is a common trope used by journalist keen on sensationalism rather than the strict truth. Misunderstandings crop up in the press. For instance, in November 2021 the BBC ran an article on their website with the headline, 'Plastic found inside seagull eggs and pellets'. This read as a little odd, since to pass into an egg this plastic would need to be incredibly small, small enough to ask if this was intrinsically a plastic or a bundle of chemicals. If the latter, how could it be identified as a plastic? Digging a little deeper, an earlier *New Scientist* article from June 2021 shed a little light, with their headline saying, 'Seagull eggs in the UK have been contaminated with plastic additives'. It appears that this headline had a closer connection to data released by the research scientists. (Note – the BBC headline has now been altered to better reflect the science.)

This casts a different angle onto the situation. It ought not detract from the point that a foreign chemical was found in the egg, but the whole language of the article sought to denigrate plastics without proof. The gulls were found to have phthalates in their system. Phthalates are found in the likes of PVC, where it is used as a plasticiser, in other words to soften the material. It is not in widespread use across all materials, rather for just a small selection of plastics and rubber. There was no indication that the gulls had taken on the phthalate from plastic, just that it was present. So where else might the gulls have picked up this chemical? Well, if they had been down at the local tip pecking at a scented candle, they could easily have imbibed this chemical. Yes, scented candles by and large contain phthalates; they help give them their strong aroma when burning. Cosmetics can contain phthalates too, as can soap, hair spray, lubricating oils, and hundreds of other product. If we are to tackle unwanted chemical ingestion we need to know what we are dealing with, heading off on a witch hunt may mean we miss other equally important areas of concern.

As an aside, the American Centres for Disease Control and Prevention (CDC) have stated further studies on phthalates are needed, but findings so far show that when entering the body they do have a tendency to break down and clear our system quickly via our urine. There are different types of phthalates, and all would need research, particularly on potential effects on the reproductive system.

Micro-materials

We have spent some time looking at microplastics in all their guises. But what of micro-ceramics, micro-metals and the myriad of other potential micro-materials? If microplastics are a concern, then surely we ought to consider the health implications of other micro-particles.

Humans can be careless creatures. Before organised removal of our waste, we tossed our rubbish wherever we could, into middens, rivers, even the street. We can all be shocked at the state of certain rivers in less developed countries, and no doubt serious action needs to be taken. However, looking at a major river of our own we can find clear evidence of our misuse of this habitat in the past. On close inspection of the Thames in central London, for instance, the extent of our litter habit may not be immediately obvious, but head down to the foreshore at low tide and look at what is underfoot. As mentioned in Chapter 7, in addition to native flint nodules, there is an expanse of red, grey and yellow debris. These indicate discarded London rubbish going back millennia. From Roman pots and tiles, to Victorian plates and dishes, and centuries of clay pipes, it is all there, washing in and out with the tide, buried and uncovered with each circuit of the moon. They gradually erode each time they are uncovered, releasing many particles of micro-ceramics, not to mention micro-glazes of lead and tin. Like plastics, these do not biodegrade either. Pots that may have started out as earthenware, once fired, are fixed, and won't easily become clay again. Glass also. It may once have started life as sand, but once formed into glass, that is how it remains until it erodes down into a sea of micro-particles. Many other materials are found on the foreshore: lengths of rope, metal clinker nails, concrete blocks... It's all there, gradually eroding over time.

Keeping London and the Thames in mind, there has for centuries been a network of metal pipes buried beneath the streets carrying water and other essential supplies. These pipes, particularly those from the Victorian era, are made primarily of cast iron and lead. They are being replaced slowly here and in other parts of the country, but for years have had water flowing through and taking up small particles which flow into our homes or into the river or sea. In effect, they have been creating micro-metals.

We now know lead to be problematic, which is why it is now banned from use in many domestic applications. It is, however, still used in other areas of the home such as flashing on the acreage of roofs traversing the country. Each time the rain pelts down, does it wash tiny particles of lead into the gutters and from there into watercourses? What about the lead that is used elsewhere, such as in shotgun pellets and fishing weights?

There has long been a call to curb the practice of using lead weights when fishing. All too often these end up snagging off a fishing line and sinking onto the sea- or riverbed, where they too can be swallowed by aquatic creatures. Some water birds for instance purposely ingest gravel to aid digestion. They can just as easily inadvertently swallow lead shot or weights – or fragments of plastic, for that matter – as the more wholesome pebbles.

Fish can also swallow gravel, and if lead is ingested what happens when we then eat that fish? May we be exposed to lead poisoning if that lead has entered the creature's bloodstream? It is possible. Fortunately, fish are less likely to be exposed to lead weights in the UK since the government imposed a ban on lead weighing between 0.06 grams and 28.35 grams (i.e. one ounce). According to the government's Freshwater Rod Fishing Rules, however, anything under or over this weight range is perfectly fine, so we aren't completely out of the woods.

Anglers aren't the only users of lead; hunters too can spread lead into the environment. Lead shot blasted into game is known to have repercussions. An article in the February 2022 edition of *Natural Geographic* reported that most bald and golden eagles in the United States suffer from lead poisoning. This poisoning was primarily found to arise from birds ingesting carcasses of animals shot with ammunition containing lead. The article went on to state that eagles are not the only wildlife suffering this poisoning, with other species affected too.

In general micro-particles have been building up for many centuries washing into the oceans, as the rain scours the land and rivers empty. We have not been concerned enough to coin a word for them. Yet more particles from disparate sources waft into the air and are breathed in as the breeze stirs up dust and other tiny smut. We could well be at risk of tying ourselves in knots if we consider all possible instances of micro-materials leaking into the atmosphere. Our everyday life causes ever greater creation of these particles. Where do we draw the line, at what point do we say, this is an acceptable amount? Can we really believe it possible to breathe air free of any kind of manmade particles?

Of course there is a difference between particles created from say the erosion of an earthenware pot and the great unknown quantity that comes attached to synthetic materials. It is easy to say that studies thus far show no harm is derived from them. Look at asbestos, a natural mineral mined for millennia and famed since the time of Ancient Greece for its ability to withstand fire. It was an essential part of any building but proved increasingly hazardous until it was finally banned in 1999. Now, of course, it is known to be the cause of serious disease, particularly where fibres have been inhaled.

This, of course, is by no means the sole source of potentially harmful particles whether of micro or nano proportions. Smoking has long been known to be damaging, with even passive smoking acknowledged to be harmful. This is because of the toxins and lethal substances carried in the smoke. Pesticides too have the potential to cause harm, as toxins in particulate form can easily be taken into the air, particularly when they are being applied to the soil. Crop dusting has the greatest potential to spread these across a wider geographical area. Some pesticides contain ingredients that are more harmful than others and can enter the atmosphere or settle on the ground. That which isn't absorbed can deteriorate by the action of sunlight or water. A raft of chemicals have use in pesticides including sulphur, boric acid and captan. Some can be acutely or chronically toxic, and calls persist to curb usage.

All manner of particulates are released from exhausts, bonfires or even the volatiles when pumping petrol, as well of course, as the aforementioned perfume or air fresheners. Larger particles can also enter the atmosphere. Each time a brake is depressed and the brake pad grinds into action particles are given off, as is the case with rusting metal flaking into the environment. The same is true of rubber in the tyres and even the road surface itself. From belching chimneys and spinning machinery to smashing crockery, micro-materials in all their guises are constantly being released into the atmosphere. Some of course will be innocuous, others are not.

There are stringent investigations into plastics in place, for the most part undertaken by academic departments with access to top-notch analytical equipment, so we'll see what may arise. Investigations and greater understanding of these materials under a whole host of conditions need to continue. It may well be that these investigations ought take on a wider remit. What is clear is that sensationalist and inaccurate media headlines do not help our understanding.

Oceans

The final focus of the book brings us right back to where we started, why the oceans are seemingly awash with plastics. Masses of waste accumulations in the ocean are apparent as never before, partly because of a natural feature of plastic. It has a relatively low density which means many have a propensity to float and so are very visible when waterborne. Just where this plastic is coming from is not always so obvious. Listening to many commentators, it seems a foregone conclusion that any plastic waste not being recycled ends up in watercourses and ultimately the sea. This notion is overly simplistic: the outcome for used plastic is multifaceted with much dependent on post-use handling. There is very understandably mounting concern over the state of the oceans, with most attention directed towards issues of pollution and its impact on the marine environment. There is absolutely no doubt that plastics are of major concern, but that does not mean we should overlook others just because we are so set against plastic. Doing so may allow other equally problematic issues to continue unaddressed. Whatever the constituent parts, it may help to establish the origins of pollution. There are two primary forms of pollution in oceans – that which is thrown in – unfortunately often purposely – and that which disgorges from rivers and watercourses. These pollution sources need plugging, and pollution which is already in the ocean dealt with as efficiently as possible. One hurdle on the path to reparation is the sheer size of the oceans and the difficulty in finding culprits where waste is purposely jettisoned.

Marine waste/pollution

Marine waste has been around for a long time and unfortunately is nothing new. Sources of waste are quite varied. Purposely discarded plastics can come from beach waste scooped by the tide and pulled seawards, or from jetsam created by maritime vessels. Unintentional waste is a result of shed cargo or flow whether from rivers or pumped seaward with discharged effluent from the land. Unsurprisingly studies are ongoing to identify sources and stem this waste. The European Parliament has issued several informative articles on the subject, either via their pressroom, the European Chemical Agency (ECHA) or the Copernicus Programme (which forms part of the European Union's Earth Observation Programme). All are useful sources of up-to-date information on the subject. Indeed, investigations by these agencies throw up some interesting facts and figures. in terms of culpability, they too cite bad waste management. One report estimates that 80% of the plastic in the ocean comes from terrestrial sources. They assess the annual plastic waste discharges from watercourses as being in the region of 1.15 to 2.41 million tons. The 80% figure incorporates all waste washing from the land, including that which enters the ocean from coastal waters. A further 20% of the plastic waste is fishing-related. Looking at this in terms of weight it comes in at a hefty 70%, with floats and buoys predominating, but ghost nets also notable, not least because of the damage they cause.

A ghost net is the term given to a fishing net that's been lost or abandoned in the ocean. According to British Sea Fishing, ghost nets tend to be gill, tangle and drift nets that are designed to hang in the sea ready to ensnare any fish swimming into them. Unsurprisingly, larger variants of these nets are environmentally damaging, as they can catch a great deal more than the targeted fish. For this reason, back in 1992 the EU imposed a ban on drift nets of 2.5km and longer (in other words nets close to two miles and longer). Nets tend to be manufactured from polypropylene or nylon, as are the fishing lines and ropes also accidently or purposely lost at sea. Depending on how measurements are taken, according to United Nations Environment Programme, ghost nets represent 46% of the so called Great Pacific Garbage Patch. Needless to say, they cause untold damage to marine life. Even if these were to be produced from natural sources, such as hemp, for instance, they will not degrade overnight. Indeed, any apparatus intended for the harsh environment of the ocean will not degrade readily. So unless current bad practices are curtailed, marine life will continue to suffer. Of course, marine plastic is not restricted to nets and fishing lines, it can also include rigid items like oil drums and fish boxes made from high density polyethylene (HDPE). One thing is clear, far too much of this waste can be sourced to the fishing industry.

Seagoing vessels ply their trade, traversing the oceans largely unnoticed or even regarded by most. Those employed in the fishing sector span the range from smaller fishing smacks to large industrial scale craft hoovering fish from the seas. Bobbing on a vast ocean, fishing vessels can appear as little more than toys on the water and so are difficult to police. When in the middle of the ocean a great deal can occur with no way of monitoring behaviour. Good practice is essential and fishermen need to take greater control and responsibility of their equipment. The problem is weighty, with estimates stating that approximately 640,000 tonnes of fishing equipment is left in oceans annually. Fishing can be the cause of often overlooked problems in the ocean.

Not wishing to point the finger solely at fishing practices, we ought to consider other shipping traffic, and here we must turn our attention to cargo ships. Again it can be argued that bad practices and shoddy management has a lot to answer for. A particular cause for concern centres on the loss of virgin plastic granules, which are primary microplastics (this term is disputed in certain areas as they are in pre-processed form, yet they are plastics all the same). These are granules destined for plastics factories, where they are processed to become those products we desire. Some of these unprocessed plastic granules are lost from ships transporting them to a manufacturer and some are lost once at their destination, where they roll or are swept into drainage systems. If due care is taken there is no need for this loss, particularly at the levels it appears to occur. Unfortunately this is by no means the only loss experienced particularly on the seas.

Cargo ships are often laden with towering stacks of containers, and, no matter what rules for securing these are in place, they lose cargo with alarming frequency. According to a survey by the World Shipping Council, covering the period of 2014 to 2017, an average 1,390 containers per year were lost; this was considered a good result, as it was nearly half that of the previous three-year period. However taking the latest available figures released in June 2022 and which quotes an average across the previous two years, this figure has increased to an annual loss of 1,629 containers. It is estimated that around 6,000 cargo ships ply the oceans with our goods with millions of containers transported in a given year. Using the figures at hand these losses work out as four or five shed containers every day. Think of these containers fastened onto a lorry and visualise how much they contain. Most are filled with the sort of goods we have become used to in our increasingly materialistic world. Needless to say, plastic is by no means the worst possible occupant of these containers, all manner of noxious chemicals or other contaminants are also shipped in this way.

There are other concerns connected with cargo ships such as their fuel source. These ships run on a form of fuel called bunker oil. Remember, right at the start we established that plastics use around 4-6% of the oil produced globally. Well, bunker oil uses another 4% of this oil. This oil tends to contain high levels of sulphur, which exhausts into the sea. Globally, shipping fleets consume in the region of four million barrels per day of high-sulphur fuel oil, with this figure expected to rise with increasing demand for Chinese goods needing shipping from the East. This is a heavy and relatively viscous fuel oil derived as a residue from crude oil. Scrubbers can be fitted to ship's exhausts to reduce their sulphur emissions. Depending on the functioning of these scrubbers, they can in themselves cause issues, as they generate washwater and residues which must be very carefully regulated; otherwise there is a risk that yet more harmful substances could be released into the seas. According to an International Council on Clean Transportation report, this is a very real risk. Not all ships have these scrubbers, as for many they would need retrofitting, a costly enterprise. Alternatively, ships' engines could be modified to run on reduced-sulphur oils, or other additives could be added to the oil to make it compatible with certain engines.

Sulphur is blamed for respiratory problems, is a component of acid rain and part of the reason for the acidification and death of vast swathes of coral reef not to mention the harm it does to other creatures. In 2020, legislation came into play limiting the sulphuric content of bunker oil bringing it from its previous limit of 3.5% to 0.5%, but many shipping commentators accept that it is largely unenforceable and may not happen. If fully adopted this limitation will have a major impact. According to the International Maritime Organisation, a 77% drop in overall sulphur oxide emissions from ships would equate to a whopping reduction of 8.5 million tonnes of sulphur oxide entering the sea each year.

Unfortunately there are more instances of the seas being used as a dumping ground. From the 1940s the sea was used as a dump for toxic nuclear waste. According to the International Atomic Energy Agency, drums of radioactive waste were tipped into the seas off California in 1946 with the practice spreading globally over the decades. These drums can hardly be said to be utterly secure as they comprised metal shells lined with tar and concrete, not impenetrable after decades in the seas. Dumping only became regulated in the 1970s, and a ban was finally declared in the 1990s. This does not mean the practice has stopped. Even now there is ongoing debate on the practice not least in Japan following the Fukushima nuclear power disaster. Questions surround their dealings with contaminated water generated from this catastrophe. According to a 2021 article in the science research journal *Nature*, more than one million tonnes of contaminated water is due to be discharged with the approval of scientists. Reportedly the water has been stored in metal containers and treated to remove radioactive material. Not everyone is convinced or happy with this judgement.

There are other oceanic sources of nuclear contaminants such as fallout from atomic testing, much of which is done at sea. Many questions arise if we are to consider this treatment of the oceans that surround us all, not least when considering just what this is doing to marine life and the fish we consume? We should not overlook plastic in the oceans but equally we should not ignore other areas of concern. Radioactivity in the water is invisible to us, but that does not mean it's not there.

No matter the source, we shouldn't turn a blind eye to any contaminants in the sea. Various campaigns and legal initiatives to protect the oceans have been employed for some time. Back in 1972, half a century ago at the time of writing, regulation of all waste being dumped in the oceans was attempted through the Convention on the Prevention of Marine Pollution by Dumping Wastes and Other Matter, also known in short as the London Convention. Its intention was to control all marine contamination and thereby to prevent ocean pollution. The fact that in the intervening fifty years the issues surrounding pollution have arguably worsened does not inspire confidence. Compliance to laws arising from this Convention have been poor, with financial backing limited and little done about enforcement. However a renewed vigour in this area has seen a number of additional proposals and investigations into the matter.

Back to shore

The most immediate source of ocean pollution is that refuse which enters the sea from coastal waters. The Mediterranean is particularly noteworthy with tourism being one of the main culprits when it comes to carelessly discarded waste entering the seas. It may be that crisp packet snatched by the wind and left to land where it may, or it may be waste 'helpfully' buried at the end of a day on the beach. Either way this refuse more often than not will be drawn into the sea. These are bad practices that can be difficult to monitor. To add to the quoted statistics, the aforementioned EU Copernicus Project states that 40% of the plastics found in the Mediterranean results from tourism in the area. Education is again key, as is a reliable method of refuse disposal. There is little use placing litter bins on the beach if they are allowed to overflow. Equally, there is little use installing open topped bins where one sharp breeze will take the contents airborne. A collective responsibility is needed both for those enjoying the beach and those tasked with the stewardship of these areas.

Heading further inland we can find other sources for ocean litter, including sewer overflows, illegal dumping and substandard industrial practices. None ought to occur and really all should be plugged. Sewer and storm drains overflowing into the sea have regulations associated with them, particularly in the UK. Unfortunately, these are often flouted. It is here that microplastics can find their way into the oceans. In fact, primary microplastics (remember, these are those purposely created plastic bits) are said to account for between 15 and 31% of microplastics in the oceans. Where these are added to scouring products, either in detergents or body exfoliates, or shed from a washing machine as clothing fibres, these end up decanting down the drain. Added to these primary sources, microplastics can also wash into a drain when eroded from road markings or car tyres or a myriad of other abraded surfaces. They can drain from the land when washed from artificial pitches, mulch covers, fertilisers or certain pesticides. Wherever the source, once in a drain there is no guarantee that this waste water will reach a treatment plant, and even when it does there is no certainty that adequate filters are in place.

While small plastics such as cigarette butts, bits of wrapper or even chewing gum (in addition to the microplastics mentioned) can potentially enter the oceans from our drains, so too can larger items like wet wipes and whatever else careless consumers chose to flush domestically. Much has been made of the pollution found in the oceans emanating from South East Asia, where bad practices and careless waste management are all too obvious. Some of this waste is purposely dumped, but much is also carried to the oceans via river courses. Ten of the worst offenders are said to contribute to 90% of the plastics entering the oceans. Eight flow through Asia and two through Africa. Four in particular are often cited and these are the Yangzi, Indus, Yellow and Hai rivers. According to a 2018 article by Scientific America, up to 2.75 million tons of plastic is decanted annually by the global network of rivers, with the Yangtze alone emptying up to 1.5 million tons of plastic into the Yellow Sea.

It must be acknowledged, before too much finger pointing, that many of these rivers possess a common factor in that they flow through countries that import plastic waste from other countries. Richer countries are very happy to wave their waste goodbye and never give it another thought, but as is apparent, the host countries have no mechanisms for dealing with the waste. This mingles with the waste they themselves produce, and vast quantities end up in watercourses and ultimately the sea. No doubt for countries struggling we ought to reconsider our practice of dumping our surplus waste on them. Helping build an infrastructure to cope with waste would certainly be beneficial to all. The seas are common to us all, so it really is in all our interest. Where countries have a little more in their coffers, such as those now reaching for the stars with their space programmes, maybe pressure needs to be brought to bear on bad practices. The news agency Reuters reported in 2019 that China dumped a total of 200.7 million cubic metres of waste into its coastal waters in 2018, which was a huge increase on the previous year. Again two major rivers were cited as the source for the majority of this waste. The Chinese Ministry of Ecology and Environment reported these figures and announced planned steps to curtail waste entering the sea. Figures refer to all waste, but it will be no surprise that much of this waste is plastic.

The Chinese authorities surveyed a portion of their ocean waste and found that it extended from the water's surface to the seabed, with plastics making up just shy of 89% of that found. The source article for this figure pointed out that China is a major player in plastics production, manufacturing some 30% of that produced globally. Of course, this doesn't mean that the Chinese are dumping all this straight into the sea; we are all responsible for the plastics in the oceans. Drilling further into the origins of waste navigating waterways towards the sea, four multinationals were associated with a large proportion of this waste. A 2021 Reuters article names PepsiCo and Coca-Cola as two of the biggest global plastics polluters. The BBC cite Nestle and Unilever coming in third and fourth place. This waste really is universal and can take the form of bottles, wrappers or other packaging. Maybe, as with the EU fishing net proposal, more could be done to encourage these multinationals to clean up.

The Chinese are trying to take a more immediate approach in this much needed clean-up with many proposals made, including limitations on plastic bags and money streamed to find a solution. This has been met with mixed success. A paper written in 2021 for an environmental journal[41] noted that China was late in taking a sustainable approach to its plastic waste, but that the steps it was now implementing were both viable and far-reaching. The article notes that the limitations on plastic bags has not done as much as it could, and is only in place for a fixed number of years; however public awareness has improved and further far-reaching laws are planned. The authors suggest that by 2025, as the national plastics pollution control plan rolls out, changes will become noticeable. As China is the world's largest plastic producer and consumer, it is to be hoped that these plans succeed.

India faces a similar problem, with highly polluted rivers causing a great deal of plastics leakage into the oceans. The United Nations Environment Programme (UNEP) helped survey the area to identify waste hotspots. The problems here are in part due to domestic habits, with waste being dumped in open areas. Clean-up drives have been undertaken by volunteers, but this really is of minimal impact given the scale of the problem. Illegal dumping, the proliferation of slums and open drains exacerbate the issue. The government, of course, is aware of concerns and has introduced a number of steps to rectify these problems. The Indian Ministry of Environment, Forest and Climate Change announced steps to phase out single-use plastic by 2022. From July 2022 items such as cotton buds with plastic sticks, plastic sticks for balloons, plastic flags, plastic cutlery, straws, trays, wrapping or packing films around sweet boxes etc. were banned. Steps were also in place to ban light weight plastic bags from December 2022 with only heavier reusable bags of 75 microns upward allowed. Measures to strengthen their infrastructure have also been announced, all of which is a very welcome step in the right direction. It is to be hoped that these two majorly populated countries really can turn a corner with their waste.

Cleaning up oceans

The UK may no longer be a member state of the EU, but the latter is still influential in the fight against plastic waste. A Marine Framework Strategy Directive exists, 'aiming to reduce plastic pollution from inland and shipping sources'. It is ultimately seeking an international legal framework to tackle marine waste. It is a worthy goal and one that hopefully will have more success than previous initiatives. There are more significant monitoring tools available now. The Copernicus Project provides helpful insights into how ocean observation could provide much needed data to begin the halt of this waste.

A concerted global effort will go a long way to avoiding yet more plastic and indeed other waste from entering the oceans. It remains to be seen what can be done with that which is there already. Larger waste such as ghost nets may physically be more straightforward to remove. As the waste becomes smaller it does throw up greater difficulties. We know that plastic can reduce to ever smaller sizes, so at what point do we say it becomes harmless? As previously mentioned plastics, or indeed any waste, reaching nano or even pico dimensions is an unknown entity in terms of potential chemical damage to either marine or human life. That naturally does not mean to say we can ignore the problem, and we should at least attempt to remove what we can. Scientists have conducted research into this area for some time, with a number of publications reporting possible breakthroughs. In several cases these involve developing plastic-eating bacteria. The first bacteria discovered to have a taste for plastic is a fussy eater, however, and will only ingest PET. As this plastic makes up so many bottles, trays and even garments globally, this might be a good start. Other modified bacteria are being investigated, with enzymes capable of 'eating' plastics being the ultimate goal.

As the Copernicus Project points outs, knowing the origin and sources of marine plastic pollution will allow improvements in waste management systems and adoption of suitable public policies. We need properly applicable policies and real action. We need to stop fixating on unrealistic targets and move the focus to where it really matters. For instance, the government face a groundswell among the voting public, saying more ought to be done to recycle plastics. Less should head to landfill, more needs recovery and reuse. Unfortunately, the reality remains that it is just not viable to recover and recycle everything. It is frustrating to hear councils won't take certain items because it doesn't suit their infrastructure, but the targets remain, and so trading ensues. Landfill must be reduced, so surplus waste is sent to other countries.

We know that this ferrying about of waste in the name of the environment is nonsensical, since far more energy is wasted in the shipping than would be recovered if recycled. Think of the amount of fossil fuel gone to making this plastic, then consider the additional fossil fuel consumed in mechanically sorting, cleaning, shredding, baling, transporting (possibly via cargo ship), then sorting, melting and reprocessing in its host country.

So let's turn everything on its head and take a look from another angle. How about having dedicated plastics landfill sites? This might be a controversial opinion, but could it work? Starting from the basics, we have certain facts to hand:

- Plastics are made mostly from finite supplies of fossil fuel
- Much of the recycling challenges relate to thin film or contaminated packaging
- It is not energy efficient to recycle this plastic either here or in a recipient country
- It is not always economical to recycle all waste plastic
- Inefficient and uneconomical waste is jettisoned for others to deal with who don't always have the appropriate mechanisms in place.

Might it be a viable option to compress this plastic waste and landfilling in dedicated sites, until such time as it becomes feasible to take positive and affirmative action, preferably in the same country? Maybe it could be a resource store easily tapped in future generations. This troublesome plastic has already been separated from the higher value plastic, so instead of transporting it to the nearest port why not head to a 'future resource store'. It could well be that we will reach a time when we have moved completely away from fossil fuels as an energy source, and plastic has become a valuable resource.

In so many ways we need to look to the future and not make a hash sweeping our problems under the carpet today.

15

CONCLUSION

We have covered a lot of ground in the chapters of this book, and hopefully along the way the merits of plastics have been highlighted, showing their utility in our bid to become better environmental citizens. There is no doubt that the subject is very complex, and an awareness of the issues has got to be useful. What is not helpful is stating all plastic should be banned simply because they are plastic, or indeed that all plastics must henceforth be plant-derived. Neither is a complete solution and neither would, on its own, save the planet.

We need an honest review of our plastics usage and avoid the frivolous use of these materials. But we also need to move away from targets that give lip service to certain sectors of society and instead look to realistic solutions. Government departments need to be cognisant of facts before imposing legislation, and for that matter so too do the media before they deride some materials and exalt others.

We all have the power to help the planet. Maybe we need to recognize this ability. We need to acknowledge there is no quick fix, we need to stop buying things for which there is no intrinsic need, and we need to make full use of the things we do buy. Embrace the vast portfolio of materials open to us, use them to their best advantage and be aware of their benefits and pitfalls. Of course it is possible to live without plastic – after all we did so for millennia – but it would mean turning the clock back in more ways than one. Obliterating plastics would mean setting humanity back a century – and just look at what we achieved in this last century, arguably more than any other preceding century. We started predominantly propelled by horse power and steam and ended with a space programme pushing to Mars and beyond. Many of those integral and manifold elements making this possible have at least some reliance on polymers.

There is no one material or fuel source that can be the panacea solving all the world's problems. Maybe we need to consider continually evolving new materials coming from a number of sources. However, these need to be understood and incorporated into the catalogue of existing materials. To ban all plastics and then wholeheartedly adopt another apparently friendly material may end up just shifting the problems we have but solve nothing in the long term.

I do hope that this book can start a discussion on how to achieve a better future with a more responsible and sustainable relationship with plastic. So are plastics just a load of rubbish? I'll leave that for you to answer.

Further Reading

Chapter 1: Why Are We So Set against Plastic: The Influence of the Media

www.opec.org/opec_web/static_files_project/media/downloads/
publications/WOO_2014.pdf

Chapter 2: A Short History – Where Did It All Start?

A Cat Turned Milk into Popular Plastic, K. Arney, Chemistry World, June, 2017

American Chemical Society National Historic Chemical Landmarks. Foundations of Polymer Science: Wallace Carothers and the Development of Nylon

Ballard D.G.H. (1986) The Discovery of Polyethylene and Its Effect on the Evolution of Polymer Science. In: Seymour R.B., Cheng T. (eds) History of Polyolefins. Chemists and Chemistry, vol 7. Springer, Dordrecht.

Sébastien Poncet, Laurie Dahlberg. The legacy of Henri Victor Regnault in the arts and sciences. International Journal of Arts and Sciences, 2011, 4 (13), pp.377-400

The Nylon Rope Trick; Journal of Chemical Education, April 1959, 36:182–184

Waldo Semon – Rubber, PVC and... bubblegum?; Claudia Flavell-While, The Chemical Engineer, IChemE

blog.sciencemuseum.org.uk/bakelite-the-first-synthetic-plastic/

www.azom.com/article.aspx?ArticleID=2101

www.historywebsite.co.uk/Museum/Tarmac/Group.htm

www.issx.org/page/EugenBaumann

www.k-online.com/en/News/February_2013

www.msthalloffame.org/patsy_sherman.htm

www.plastiquarian.com

www.qz.com/785119/the-forgotten-tropical-tree-sap-that-set-off-a-victorian-tech-boom-and-gave-us-global-telecommunications/

www.sciencehistory.org/historical-profile/karl-ziegler-and-giulio-natta

www.sciencehistory.org/historical-profile/stephanie-l-kwolek

Chapter 3: Categorising These Materials – the Sciency Bit

omnexus.specialchem.com/selection-guide/polypropylene-pp-plastic

www.ncbi.nlm.nih.gov/pmc/articles/PMC2873014/

Chapter 4: The Language around Plastics, from Media to Marketers

theconversation.com/the-world-of-plastics-in-numbers-100291

Chapter 5: To BPA or Not to BPA

Comprehensive Reviews in Food Science and Food Safety, Vol. 12, 2013

Cure mechanisms of diglycidyl ether of bisphenol A (DGEBA) epoxy with diethanolamine; Polymer, Volume 105, 22 November 2016, Pages 243-254

Epoxy Resin Committee-July 2015; epoxy resins assessment of potential BPA emissions–summary paper:

Fundamentals of Microbiology: Body Systems Edition, 2014, Jeffrey C. Pommerville, Jones and Bartlett Publishers, Inc

Natural occurrence of bisphenol F in mustard; Food Additives & Contaminants: Part A, Volume 33, 2016 - Issue 1, Pages 137-146

www.bbc.co.uk/news/uk-49280709

www.chemicalsafetyfacts.org/bpa-bisphenol-a/

www.edu.rsc.org/download?ac=14902

www.efsa.europa.eu/en/topics/topic/bisphenol Bisphenol A and replacements in thermal paper: A review, Maria K Björnsdotter et al, Chemosphere, 2017 Sep

www.epoxy-europe.eu/safety/

www.epoxy-europe.eu/wp-content/uploads/2016/09/epoxy_erc_bpa_whitepapers_summarypaper.pdf

www.factor.niehs.nih.gov/2018/10/science-highlights/clarity_bpa/index.htm

www.factsaboutbpa.org/benefits-applications/why-bpa

www.food.gov.uk/safety-hygiene/bpa-in-plastic

www.food.gov.uk/safety-hygiene/bpa-in-plastic

www.guichon-valves.com/faqs/epoxy-resins-manufacturing-process-of-epoxy-resins/

www.ncbi.nlm.nih.gov/pmc/articles/PMC4206219/

www.ncbi.nlm.nih.gov/pubmed/20077198

www.newsletter.echa.europa.eu/home/-/newsletter/entry/moving-away-from-bpa-in-thermal-paper

www.onlinelibrary.wiley.com/doi/pdf/10.1111/1541-4337.12028

www.onlinelibrary.wiley.com/doi/pdf/10.1111/1541-4337.12028

www.polymerdatabase.com/home.html

www.pubmed.ncbi.nlm.nih.gov/31055636/

www.sciencedaily.com/releases/2008/01/080130092108.htm

www.scientificamerican.com/article/plastic-not-fantastic-with-bisphenol-a/

www2.mst.dk/udgiv/publications/2011/04/978-87-92708-93-9.pdf

www2.mst.dk/Udgiv/publications/2015/05/978-87-93352-24-7.pdf

Chapter 6: When Is a Bioplastic Not a Bioplastic?

Advanced Industrial and Engineering Polymer Research; Volume 3, Issue 2, April 2020, Pages 60-70

Bio-Based Materials For Use In Food Contact Applications;

Fera project number FR/001658, Report to the Food Standards Agency, June 2019

Biodegradable and Compostable Plastics: A Critical Perspective on the Dawn of their Global Adoption; Dr. Rosaria Ciriminna, Dr. Mario Pagliaro, Chemistry Open. 2020 Jan; 9(1): 8–13.

Biodegradable and Compostable Plastics: A Critical Perspective on the Dawn of their Global Adoption; Chemistry Open. 2020 Jan; 9(1): 8–13.

blogs.ei.columbia.edu/2017/12/13/the-truth-about-bioplastics/

Can Polyhydroxyalkanoates Be Produced Efficiently From Waste Plant and Animal Oils?, Arthy Surendran et al, Frontiers in Bioengineering and Biotechnology, March 2020

Environmental Science and Technology, 53, 9, 4775–4783, April 28, 2019: Environmental Deterioration of Biodegradable, Oxo-biodegradable, Compostable, and Conventional Plastic Carrier Bags in the Sea, Soil, and Open-Air Over a 3-Year Period

Henry Thurber & Greg W. Curtzwiler (2020); Suitability of poly(butylene succinate) as a coating for paperboard convenience food packaging, International Journal of Biobased Plastics,

Home Composting; Factsheet April 2015, European Bioplastics

Plastics to Energy: Fuel, Chemicals, and Sustainability Implications; Elsevier Inc., Adriaan S. Luyt, Sarah S. Malik, 2019

Polymer Properties Database, Chemical Retrieval on the Web (CROW)

Royal Society of Chemistry; Sustainable Plastics – the Role of Chemistry, Summary of a roundtable discussion meeting hosted by the RSC Materials Chemistry Division, March 2019

Rubber & Plastics News; Oct. 8, 2019

Understanding plastic packaging and the language we use to describe it; Wrap, www.wrap.org.uk

www.bioplasticsnews.com/2019/04/13/what-is-the-difference-between-biodegradable-compostable-and-oxo-degradable/

www.docs.european-bioplastics.org/publications/fs/EUBP_FS_
Standards.pdf

www.frontiersin.org/articles/10.3389/fbioe.2020.00169/full

www.interpack.com/en/TIGHTLY_PACKED/SECTORS/
BEVERAGES_PACKAGING/News/PEF_%E2%80%93_100_
natural,_100_recycled,_100%C2%A0_material_benefits

www.polymerdatabase.com/polymer%20classes/Bioplastics.html

www.recyclenow.com/recycling-knowledge/
packaging-symbols-explained

www.rsc.org/globalassets/04-campaigning-outreach/policy/envi-
ronment-health-safety-policy/plastics-sustainability.pdf

www.sulzer.com

www.wrap.org.uk

Chapter 7: Packaging

www.americanchemistry.com/chemistry-in-america/news-trends
/blog-post/2018/there-s-a-reason-we-use-plastics-to-package-food

www.fcrn.org.uk/sites/default/files/Fruitnveg_paper_2006.pdf

www.multibriefs.com/briefs/exclusive/barrier_packaging_2.
html#.Xw36O-fTXIU

www.newfoodmagazine.com/article/1893/
solubility-of-carbon-dioxide-in-meat/

www.ourworldindata.org/faq-on-plastics

www.ourworldindata.org/faq-on-plastics

www.ourworldindata.org/plastic-pollution#how-much-of-ocean-
plastics-come-from-land-and-marine-sources

www.plasticseurope.org/

www.polymerdatabase.com/Films/Multilayer%20Films.html

www.polymerinnovationblog.com/wp-content/uploads/2017/03/
tr-guide-multilayers-v6.pdf

www.resource.co/article/agricultural-plastics-collection-scheme-
launched

www.scientistlive.com/content/
gas-mixtures-help-preserve-quality-packaged-meats
www.wrap.org.uk/content/plastic

Chapter 8: Construction and Building

Prospect of bamboo as a renewable textile fiber, historical overview, labeling, controversies and regulation, L. Nayak et al, Fashion and Textiles, Article number: 2 2016
www.bpf.co.uk
www.carltonservices.co.uk/news/how-is-a-refrigerator-made/
/collection.sciencemuseumgroup.org.uk
www.contrado.co.uk/blog/what-is-rayon/
www.core.ac.uk/download/pdf/194313677.pdf
www.designingbuildings.co.uk/
www.essentialchemicalindustry.org/polymers/polyamides.html
www.essentialchemicalindustry.org/polymers/polyesters.html
www.foamtechchina.com/eva-foam-material/
www.hydratech.co.uk/Technical/Propylene-Glycols/0/59
www.jdpipes.co.uk/knowledge/mains-supply/colours-and-uses.
html
www.sciencedirect.com/topics/chemistry/microfiber
www.sewport.com/fabrics-directory/nylon-fabric
www.theflooringgroup.co.uk/what-is-linoleum/
www.worldconstructiontoday.com/articles/
materials-used-in-the-construction-of-smart-buildings/

Chapter 9: Electrical/Electronic

41st Risø International Symposium on Materials Science, IOP Conf. Series: Materials Science and Engineering 942 (2020) 012015
Automotive World; Risky business: the hidden costs of EV battery raw materials, Nathan Picarsic, November 23, 2020

Energy & Environmental Science 8.8 (2015); Wang et al; 2250-2282

Materials for Wind Turbine Blades: An Overview; Leon Mishnaevsky et al, Materials (Basel). 2017 Nov; 10(11): 1285.

Natural History Musuem; Lithium carbonate has been produced from UK rocks for the first time, Josh Davis, 19 January 2021

Nature; Electric cars and batteries: how will the world produce enough? Davide Castelvecchi, 17 August 2021

omnexus.specialchem.com

Production, use, and fate of all plastics ever made; R. Geyer et al, July 2017; Science Advances 3(7):e1700782

Thacker Pass Lithium Mine Project; Final Environmental Impact Statement; DOI-BLM -NV -W010-2020-0012 –EIS; December 4, 2020

www.bbc.co.uk/news/business-51325101

www.bpf.co.uk

www.cait.rutgers.edu/research/
piezoelectric-energy-harvesting-in-airport-pavement/

www.cisco.com/c/en_uk/solutions/data-center-virtualization/
what-is-a-data-center.html

www.designlife-cycle.com

www.euronews.com/green/2021/06/18/
old-wind-turbines-are-being-reborn-as-bridges-in-ireland

www.hydroreview.com/world-regions/how-composite-materials-can-be-used-for-small-hydro-turbines/#gref

www.madehow.com/Volume-1/Wind-Turbine.html

www.maritime-executive.com/article/new-world-s-largest-wind-turbine-as-offshore-wind-scale-up-continues

www.ncbi.nlm.nih.gov/pmc/articles/PMC5706232/

www.plenco.com

www.power-technology.com/features/featurepower-from-foot-steps-london-trials-intelligent-streets-4882232/

www.roadsbridges.com/new-aspect-roadways

www.semprius.com/wind-turbine-blades-size/
www.usgs.gov/faqs

Chapter 10: Transport

Airline weight reduction to save fuel: The crazy ways airlines save weight on planes; Hugh Morris; The Telegraph, London, Sep 4 2018

cen.acs.org/articles/95/i45/Plastics-makers-plot-future-car.html

Chemical & Engineering News ; Plastics makers plot the future of the car; Alexander H. Tullo, November 13, 2017; Volume 95, Issue 45

HANSARD 1803–2005: 7 February 1934, Commons Sitting

Nitin Girdhar Shinde and Dilip Mangesh Patel 2020 IOP Conf. Ser.: Mater. Sci. Eng. 810 012033

Usage of polymeric fuel tanks in the automotive industry; L N Shafigullin et al 2018 IOP Conf. Ser.: Mater. Sci. Eng. 412 012071

WRAP; The Composition of a Tyre: Typical Components, Project code: TYR0009-02

www.americanchemistry.com/better-pol-icy-regulation/transportation-infrastructure/corporate-average-fuel-economy-cafe-emissions-compliance

www.automotiveplastics.com/wp-content/uploads/FuelTank.pdf

www.automotiveplastics.com/wp-content/uploads/Transitioning-to-a-Circular-Economy_10-1-20_singlepage.pdf

www.bbc.co.uk/ahistoryoftheworld/

www.boating.guide/how-canoes-are-made/

www.continental-tyres.co.uk/car/all-about-tyres/tyre-essentials/tyre-mixture

www.dunlop.eu/en_gb/consumer/learn/how-tires-are-made.html

www.historic-uk.com/HistoryMagazine/DestinationsUK/French-Cannons-as-Street-Bollards/

www.resource-innovations.com/resources/

Chapter 11: Good to Be Green?

Aarhaug, T.A., Ratvik, A.P. Aluminium Primary Production Off-Gas Composition and Emissions: An Overview. JOM 71, 2966–2977 (2019).

An Overview of Hydrotreating, Emmanuel Ortega, CEP, October 2021

BEIS Industrial Fuel Switching Phase 2: Alternative Fuel Switching Technologies

Best Ways to Cut Carbon Emissions from the Cement Industry Explored, Caroline Brogan, Imperial College News, 20 May 2021

Environmental assessment of water, chemicals and effluents in leather post-tanning process: A review; E. Hanson et al; Environmental Impact Assessment Review; Volume 89, July 2021

Environmentally Benign Approaches for Pulp Bleaching (Second Edition), Pratima Bajpai , 2012

Environmentally Benign Approaches for Pulp Bleaching; Pratima Bajpai (Second Edition), 2012

for the Glass Sector, November 2019

Introduction to cement and concrete, Peter A. Claisse, Civil Engineering Materials, 2016

joint-research-centre.ec.europa.eu/jrc-news-and-updates/eu-climate-targets-how-decarbonise-steel-industry-2022-06-15_en

news.un.org/en/story/2021/10/1102432

Plastic bottles vs. aluminium cans: who'll win the global water fight?; Eric Onstad, October 2019, Reuters Commodities News

Plastics go Green, Cynthia Washam; Chemmatters, AprilL 2010

Power Generation, Woodhead Publishing Series in Energy, 2019, Pages 83-97Chapter 7

Pulp and Paper Production Processes and Energy Overview, Pratima Bajpai,; Pulp and Paper Industry, 2016

Sustainability of Municipal Solid Waste Management; M.Salah

et al. Sustainable Industrial Design and Waste Management,; 2007

Sustainable Commodities Marketplace Series 2020; Global Market Report: Cotton,

UNESCO: Value of Water Research Report Series No. 18; The water footprint of cotton consumption, A.K. Chapagain et al; September 2005

Vanilla is a forest industry by-product ; Julie Crick, Michigan State University, January 2017

www.bbc.co.uk/news/business-56716859

www.britishsteel.co.uk/what-we-do/how-we-make-steel/

www.capp.ca/oil/extraction/

www.cfda.com/resources/materials/detail/silk

www.chemguide.co.uk/inorganic/extraction/aluminium.html

www.copperalliance.org.uk

www.cotton.org/

www.cottonaustralia.com.au/

www.cottonworks.com/wp-content/uploads/2018/01/Dyeing_Booklet.pdf

www.eia.gov/environment/emissions/co2_vol_mass.php

www.engineeringenotes.com/engineering/glass/glass-composition-properties-types-and-treatment-materials-engineering/46795

www.european-aluminium.eu/

www.fertilizerseurope.com/fertilizers-in-europe/how-fertilizers-are-made/

www.glassallianceeurope.eu/en/what-is-glass

www.glass-sellers.co.uk/our-industry/industrial-glass/

www.historic-uk.com/HistoryUK/HistoryofBritain/Cotton-Industry/

www.historyofglass.com/glass-making-process/glass-ingredients/

www.imperial.ac.uk/news/232638/

low-cost-intelligent-soil-sensors-could-help/

www.iwto.org/sustainability/

www.madehow.com/Volume-6/Cotton.html

www.mylearning.org/stories/originating-in-leeds/122

www.nachi.org/history-of-concrete.htm

www.nextnature.net/story/2019/how-to-biofabricate-leather

www.nsmedicaldevices.com/news/uk-covid-19-vaccine-bd/

www.oxfordreference.com/view/10.1093/oi/
    authority.20110803095502737

www.paper.org.uk/CPI/Content/Information/Papermaking-
    Process.aspx

www.pilkington.com/

www.reuters.com/article/us-environment-plastic-aluminium-
    insight-idUSKBN1WW0J5

www.reuters.com/world/uk/uk-pay-tens-millions-get-carbon-
    dioxide-pumping-again-2021-09-22/

www.science.howstuffworks.com/plastic6.htm

www.sciencelearn.org.nz/image_maps/41-wool-processing-fleece-
    to-fabric

www.theconversation.com/oil-companies-are-going-all-in-
    on-petrochemicals-and-green-chemistry-needs-help-to-com-
    pete-153598

www.theworldcounts.com/

www.usgs.gov/centers/nmic/copper-statistics-and-information

Chapter 12: Then and Now: Some better Uses and Some Hidden Uses

Advances in the manufacture of sausage casings, Z. Savic,
    Advances in Meat, Poultry and Seafood Packaging, 2012

Applications of Bioactive Seaweed Substances in Functional Food
    Products, Yimin Qin, Bioactive Seaweeds for Food Applications,
    2018

Current Archaeology, Peasant houses in Midland England, May 1, 2013

Homecraft and Homemaking, Margeret Nicol, McDougall's Educational Co Ltd, 1940

www.archaeology.co.uk/articles/peasant-houses-in-midland-england.htm

www.bankofengland.co.uk/

www.carbonbrief.org/in-depth-qa-how-will-tree-planting-help-the-uk-meet-its-climate-goals

www.fisheries.noaa.gov/

www.instituteofmaking.org.uk/materials-library/material/dish-sponge

www.pnas.org/content/early/2020/02/04/1910485117

www.polyurethanes.org/ where-is-it/footwear/

www.regencyredingote.wordpress.com/2011/11/11/sand-a-regency-cleaning-agent/

www.sciencedirect.com/topics/food-science/sausage-casing

www.silicone.co.uk/news/an-introduction-to-silicone/

www.tea.co.uk

www.theguardian.com/travel/2013/jul/27/history-of-englands-forests

www.toysoldierco.com/resources/toysoldierhistory.htm

www.vam.ac.uk/articles/the-history-of-hand-knitting

www.wonderopolis.org/wonder/what-makes-glitter-sparkle

13. Go Compare – Choosing Plastic or an Alternative

An updated lifecycle assessment study for disposable and reusable nappies; Science Report – SC010018/SR2; Published by the Environment Agency; 2008

Comparative Life Cycle Assessment of an Artificial Christmas Tree and a Natural Christmas Tree; American Christmas Tree Association; Thinkstep AG; November 2010

Life Cycle Assessment of grocery carrier bags Environmental Project no. 1985, : The Danish Environmental Protection Agency; 2018

Life cycle assessment of supermarket carrier bags: a review of the bags available in 2006 - Report: SC030148; Published by the Environment Agency; 2011

Materials World; December Issue 2019

www.climateandweather.net/global-warming/deforestation.html

14. Cleaning Up Our Act

Deposit- and suspension-feeding sea cucumbers (Echinodermata) ingest plastic fragments, E.R. Graham, J.T. Thompson, Journal of Experimental Marine Biology and Ecology, 368 (2009), pp. 22-29

Environmental implications of plastic debris in marine settings—entanglement, ingestion, smothering, hangers-on, hitch-hiking and alien invasions; Murray R. Gregory, Philos Trans R Soc Lond B Biol Sci. 2009 Jul 27; 364(1526): 2013–2025.

marine.copernicus.eu/explainers/phenomena-threats/ plastic-pollution/from-plastic-marine-pollution

Most U.S. eagles suffer from lead poisoning, study suggests, Douglas Main; National Geographic, 17 February 2022

The Value of China's Legislation on Plastic Pollution Prevention in 2020, Liu, J., Yang, Y., An, L. et al; Bull Environ Contam Toxicol ; 2021

www.britishseafishing.co.uk/commercial-fishing-methods/

www.cdc.gov/biomonitoring/Phthalates_FactSheet.html

www.dw.com/en/fukushima-how-the-ocean-became-a-dumping-ground-for-radioactive-waste/a-52710277

www.echa.europa.eu/documents/10162/ db081bde-ea3e-ab53-3135-8aaffe66docb

www.efsa.europa.eu/sites/default/files/2021-05/7.human-expo-sure-micro-nanoplastics-citizens-concerns-pahl.pdf

www.environmentagency.blog.gov.uk/2020/07/02/
combined-sewer-overflows-explained/
www.europarl.europa.eu/news/en/head-
lines/society/20181116STO19217/
microplastics-sources-effects-and-solutions
www.gov.uk/freshwater-rod-fishing-rules/tackle-you-can-use
www.iaea.org/sites/default/files/31404684750.pdf
www.imo.org/
www.iumi.com/news/iumi-eye-newsletter-march-2021
www.nature.com/articles/d41586-021-01225-2
www.newsecuritybeat.org/2019/08/
marvelous-plastic-waste-chinas-endgame/
www.pan-uk.org/health-effects-of-pesticides/
www.pib.gov.in/PressReleaseIframePage.aspx?PRID=1745433
www.reuters.com/article/us-china-pollution-oceans-idUSK-
BN1X80FL
www.reuters.com/business/retail-consumer/
pepsico-slash-plastic-use-sustainability-push-2021-09-15/
www.sciencedirect.com/science/article/pii/
S0025326X11005133#b0135
www.sciencedirect.com/science/article/pii/S0304389419301979
www.scientificamerican.com/article/microplastics-earth-has-a-
hidden-plastic-problem-mdash-scientists-are-hunting-it-down/
www.scientificamerican.com/article/stemming-the-plastic-tide-
10-rivers-contribute-most-of-the-plastic-in-the-oceans/
www.theicct.org/publications/
global-scrubber-discharges-Apr2021
www.worldshipping.org

Endnotes

1 Plastic Materials, John Brydson, Butterworth Press. It is still in print and is in its eighth edition.

2 Polymer is the generic name for plastics and rubbers.

3 A somewhat disparaging article regarding gutta percha appeared in Volume 50 of, The Mechanics' Magazine and Journal of Science, Arts, and Manufactures, printed in 1849. It considered the novel Irish usage of gutta-percha as a means for making the seals on their official documents. The article shows the writers of the time to be amazed that the Irish should find such a good use for gutta-percha.

4 https://www.chemicalsafetyfacts.org/bpa-bisphenol-a/

5 Natural occurrence of bisphenol F in mustard; Food Additives & Contaminants: Part A, Volume 33, 2016 - Issue 1, Pages 137-146

6 Arch Toxicol. 2019 Jun;93(6):1485-1490. doi: 10.1007/s00204-019-02442-5. Epub 2019 May 5.

7 https://epoxy-europe.eu

8 https://www2.mst.dk/udgiv/publications/2011/04/978-87-92708-93-9.pdf

9 https://echa.europa.eu/hot-topics/bisphenol-a

10 Chemical study on Bisphenol A, Report: RIKZ/2001.027, Rijkswaterstaat Institute for Coastal and Marine Management (RIKZ)

11 A very interesting review on the migration of chemicals from polymer food packaging was conducted by the Institute of Food Technologists and if you are interested in learning more on this subject, it is worth a read. It can be found at: https://onlinelibrary.wiley.com/doi/pdf/10.1111/1541-4337.12028

12 www.food.gov.uk/sites/default/files/media/document/bio-based-materials-for-use-in-food-contact-applications.pdf

13 Biodegradable Mulch; Report No. EXT-2018-01, June 2018, Douglas

G. Hayes and Markus Flury

14 Standards for Bio-Based, Biodegradable, and Compostable Plastics, Call for Evidence, 2019, DEFRA

15 https://blogs.ei.columbia.edu/2017/12/13/the-truth-about-bioplastics/

16 Standards for Bio-Based, Biodegradable, and Compostable Plastics, Call for Evidence, 2019, DEFRA

17 www.wrap.org.uk

18 https://docs.european-bioplastics.org/publications/bp/EUBP_BP_Home_composting.pdf

19 Environmental Science and Technology, p. 4775–4783, April 28, 2019

20 Deutsche Welle (DW): Stiftung Warentest foundation

21 https://english.nvwa.nl/documents/consumers/food/safety/documents/advice-from-buro-on-the-health-risks-of-bamboo-cups.

22 http://www.flanderstoday.eu/

23 Plastics Europe

24 www.davpack.co.uk/retail-packaging

25 https://news.un.org/en/story/2013/09/448652

26 www.wrap.org.tuk/content/plastic

27 https://www.gov.uk/guidance/check-if-you-need-to-register-for-plastic-packaging-tax

28 www.bpf.co.uk

29 collection.sciencemuseumgroup.org.uk

30 Power Technology, Analysis, Updated February 6th, 2020.

31 Materials for Wind Turbine Blades: An Overview; Leon Mishnaevsky et al, Materials (Basel). 2017 Nov; 10(11): 1285

32 41st Risø International Symposium on Materials Science, IOP Conf. Series: Materials Science and Engineering 942 (2020) 01201

33 www.euronews.com/green/2021/06/18/old-wind-turbines-are-being-reborn-as-bridges-in-ireland

34 www.wrap.org.uk

35 Home Craft and Homemaking published by McDougall's Educational Co Ltd.

36 Allergenic Ingredients in Personal Hygiene Wet Wipes from Dermatitis. 2017 Sep/Oct;28(5):317-322

37 www.thames21.org.uk

38 European Di Isocyanate and Polyol Producers Association.

39 An updated lifecycle assessment study for disposable and reusable nappies; Science Report – SC010018/SR2; Published by the Environment Agency; 2008, https://assets.publishing.service.gov.uk/government/uploads/system/uploads/attachment_data/file/291130/schoo808boir-e-e.pdf

40 www.fertilizerseurope.com/circular-economy/micro-plastics/

41 The Value of China's Legislation on Plastic Pollution Prevention in 2020, Liu, J., Yang, Y., An, L. et al; Bulletin of Environmental Contamination and Toxicology; September 2021

Acknowledgements

I would like first and foremost to thank all at Hero Press who believed as I did that this side of plastic's place in the world needed telling. Without your guidance and support this book would not exist. I would like to extent deepest thanks to Christian Müller who kept my writing on the right path and who skilfully shepherded me away from setting off down too many side roads. He shaped the book to what you see here. All facts have been carefully checked and any errors are my own.

I am indebted to the many researchers, academics and agencies across a multitude of disciplines who have made their findings public. Without their open publishing I could not have written with authority.

Deepest thanks are due to my family who have put up with me scribbling away at this book for the last few years. Above all special thanks goes to my son Sebastian who read parts of early drafts and provided many helpful suggestions and useful input. You have wisdom beyond your years and I hope as you go out into the world you can enjoy a more sustainable planet.